PROGRAMMABLE PLANET

PROGRAMMABLE PLANET

THE SYNTHETIC BIOLOGY REVOLUTION

TED ANTON

Columbia University Press   *New York*

Columbia University Press
Publishers Since 1893
New York   Chichester, West Sussex
cup.columbia.edu

Library of Congress Cataloging-in-Publication Data
Names: Anton, Ted, author.
Title: Programmable planet : the synthetic biology revolution /
Ted Anton.
Description: New York : Columbia University Press, [2023] |
Includes bibliographical references and index. |
Identifiers: LCCN 2022051472 |
ISBN 9780231205108 (hardback) | ISBN
9780231555845 (ebook)
Subjects: LCSH: Synthetic biology.
Classification: LCC TP248.27.S95 A58 2023 |
DDC 660.6—dc23/eng/20230311
LC record available at https://lccn.loc.gov/2022051472

Cover design: Mary Ann Smith
Cover image: Shutterstock

For my granddaughter and her generation, in the hope that some of this science may help lead to a better world

So, over that art, which you say adds to nature, is an art that nature makes.

—William Shakespeare, *The Winter's Tale*

CONTENTS

III BIOINDUSTRIAL REVOLUTION

PROGRAMMABLE PLANET

INTRODUCTION

I n the squat labyrinth of Latimer Hall at the University of California, Berkeley, twenty-seven-year-old Emily Hartman had been trying for two years to understand the building of a virus shell. This winter afternoon in February 2018, tired and stressed, she glanced at the Petri dishes.

The team she worked with was seeking to unlock the ways virus shells helped killers infiltrate a healthy body. The virus slipped through a cell's defenses, the shell dissolved, and the invader delivered its deadly payload. The plan was to transform viruses from killers into delivery vehicles for medicines and vaccines to attack diseases. Every molecule in the virus shell had its role, and understanding how those molecules were assembled could reveal critical clues to a whole new way of saving lives, or so her team hoped.

Two thousand miles away her midwestern supervisor, thirty-eight-year-old Danielle Tullman-Ercek, an associate professor of biological engineering at Northwestern University, waited in a freezing conference room overlooking Lake Michigan. Dark-haired and self-possessed, Tullman-Ercek had grown up all over the country as an air force child. She spent her earliest years in upstate New York, only remembering being cold and waiting for her father to return from work as a B-52 navigator. Running a lab, she sometimes felt, was like navigating a plane through the clouds, at night, in the rain.

Tullman-Ercek had conducted research at the University of California, San Francisco, with a pioneer of a new field called synthetic biology, the science of engineering life at the level of the cell. In nearby Berkeley, a community of tinkerers and dreamers gathered in the mountains above the city, talking about customizing cells to make food and fuels, medicines and clothing, perhaps to reveal life's origin or resurrect the animals and plants of the past. The small group of computer and chemical engineers, molecular and plant biologists, and others met at the new Synthetic Biology Research Center in Emeryville, California. Some spun off companies to make vaccines, cancer treatments, cosmetics, and new energy sources. Tullman-Ercek felt they were a group of geniuses few in the world knew about.[1]

But delivery on the promises was slow. Now she had her own lab. The pressures were intense. Her team worked to understand how a virus shell assembled itself and how the tactics used by viruses could be turned into medicines. As her team waited for the video link-up, Tullman-Ercek looked out over Lake Michigan's frigid expanse.

From the Berkeley lab, Emily Hartman's face came into view. She offered her data on a split video screen, point by point, to the group in the Illinois conference room. What they heard stunned them. Listening while poring through the data as she talked, they learned that Hartman had replaced each of a possible twenty amino acids, one at a time, at 129 sites on the virus shell, revealing some 2,600 ways a virus could be redesigned. A series of quiet *wow*'s followed. They saw how easily the virus could be manipulated, producing more exclamations. "We weren't very eloquent," recalled Tullman-Ercek. "We should have yelled, Eureka!"

All Hartman heard was silence. At first, she thought she had done something terribly wrong.[2] But what Tullman-Ercek saw was the potential for using viruses as treatments, turning killers into healers. She stared. In that moment, it seemed, the dream might finally make good on its promise.

This book tells the story of a new science, engineering life forms to make what humans need—better medicines, healthier foods, sustainable materials and energy sources, and new ways to remediate our planet. This science-fiction-like tinkering promises to help confront some of our biggest problems, from reducing our dependence on fossil fuels to fighting new viruses, confronting climate change to recycling industrial pollution. Its challenge and danger is the potential for mistakes in designing living organisms, from long-extinct species to healthier humans, and the role of money in the efforts to regulate new technologies or make them available to all.

When you eat cheese, certain fruits or vegetables, sleep in or clean your cotton sheets, or have your morning cereal or evening dinner, you may be consuming the products of synthetic biology or genetic engineering. Today, the pace of progress in discovery is accelerating rapidly. The crises of emerging or drug-resistant diseases and environmental collapse are inspiring a race to transform the manufacturing of materials that clothe, feed, heal, and house us. Then came COVID-19, and these small groups of researchers jumped to the forefront of a global crisis in a way that no one had predicted.

Almost anything in the cell, its DNA or RNA or proteins, can be customized or improved. Bacteria and yeast can be engineered to make insulin for people with diabetes, protein for meatless burgers, or fibers for clothing materials. Scientists can create DNA, edit genes, and transfer cell pathways from one organism to another without the normal brakes put on by biology. The new techniques feature automated handlers, gene sequencers, gene editors and manufacturers, sound-guided experiments running twenty-four hours a day, and more. Automating evolution, synthetic biology offers a promise of industrial revolution similar to that of the silicon chip or the automobile assembly line. Researchers are competing to attack global crises, but they could potentially unleash unintended social effects, raising some of the biggest ethical challenges of our time.

There is nothing new about modifying life to make products that humans need. Efforts to manipulate biology date back to the beginning of agriculture and the first wines, beers, and cheese, as humans began breeding dogs, horses, goats, sheep, and cows. Modern synthetic biology had roots in twentieth-century wartime chemistry. England's Chaim Weizmann tapped bacteria to ferment sugars into chemicals for gunpowder during World War I,[3] and American researchers industrialized penicillin production from fermentation tanks during World War II. In the 1970s, modified bacteria and yeast made insulin, inspiring the first wave of biotechnology promises.

What is new is the speed, scale, and power of researchers' ability to remake life. Today, hundreds of labs are fine-tuning gene pathways and DNA, and gene editing is taking place in homes, garages, and backwoods stables for cattle and horses. In agriculture, many of the tomatoes, apples, and oranges we eat are engineered to resist disease or to enhance their flavor or shelf life. Most cheese and many cold water detergents are made using enzymes produced by modified organisms. The same holds for many vegetarian meats. Some of the more whimsical quests seem drawn from the pages of science fiction—glowing trees to light highways, designer plants to cool Earth, and cells to detect and attack disease.

For twenty years, such promises outpaced reality. Biological information was mostly inseparable from the living molecules that processed it. It took an incredibly long time to understand the steps of the simplest metabolic pathway that might produce a useful enzyme. Once scientists understood the steps, they had to coax industrial production from reluctant living organisms.

Those steps, however, need not be fully understood to be manipulated. The first altered organisms were simple bacteria and yeast, modified by techniques like gene transfer, moving a gene from one organism to another; gene synthesis, splicing DNA snippets in a lab; and gene editing, changing life's instructions like a musician changes key. Today, researchers modify plants and animals to make them more productive and some are in trials with humans at risk for disease.

One key early success was in medicine. The California company Amyris made a malaria drug that was subsequently shipped to millions of people affected by the disease in Africa.[4] Other companies make everything from sustainable foods and medicine to clothing, cosmetics, and fuel. They competed for development money as they attacked cancer tumors, confronted global warming, and made fabrics and building materials, safer chemicals, and renewable energy sources.

There are many definitions of synthetic biology, but for us it begins with modifying organisms by changing their DNA. That was the gene engineering revolution of the 1970s. By the 2000s, several techniques improved so rapidly they made it possible to modify life on a grander scale, leading to the "construction of new biological entities such as enzymes, genetic circuits, and cells to make useful products," in the words of the Engineering Biology Research Consortium (EBRC). This unique community of researchers started with processes similar to those used by a microbrewery to make beer and wine in fermentation vats and then expanded and extended them. In this book, I define synthetic biology broadly to include techniques like gene editing and directing the evolution of life's capabilities.

Synthetic biology began in departments of biology, engineering, physics, and chemistry, in California, Massachusetts, and elsewhere. The researchers included outcasts and visionaries who tried making circuits in cells or changing cell functions and structures altogether. At first they had little idea what changes would be most beneficial, but with experimentation, cheaper DNA, and speedier cloning, they programmed useful abilities in computer models and built and tested them in the lab. They wanted to make biology into an information science, the information coming from molecular treasure troves billions of years old. Nature had long created the materials—wood, fur, meat, skin, bones, petroleum—that made our food, housing, medicines, fuel, and clothing. Could science go further and improve on life's original production?

They worked in small labs, using painstaking techniques, seeking recognition and consensus without a common language. Gradually, they agreed on standards, building the power to reproduce one another's experiments, and developed a philosophical framework for the creation of altered life. Some argued the new science must be democratic, its technologies so inexpensive they could be available to anyone.

With COVID-19, engineering genetic material reached star status. Moderna and BioNTech used the new science to fashion vaccines from mRNA, life's messenger, beating out some 180 other vaccine candidates. The mRNA vaccines alerted the body to produce its own defenses, making a platform now being used to attack new diseases. Using synthetic biology in other ways, other vaccines tapped DNA to make the human body the vaccine maker. Suddenly, world organizations, national governments, and private foundations poured in money. Synthetic biology played a leading role in what one researcher called "the greatest science experiment in vaccinology that's ever been done."[5]

To some readers, altering life forms sounds like the plot of *Frankenstein* or *Jurassic Park*. A science that started with creativity and idealism was indeed driven in part by those seeking fame or profit. For most, however, the goal was to solve problems—confronting a pandemic, climate change, resistant diseases, pollution, the need for new energy sources, and the disposal of nuclear waste. This book helps readers understand and evaluate critical research as it took center stage.

This could be synthetic biology's big moment. Claiming the world is "on the cusp of an industrial revolution powered by biotechnology," a 2022 White House Executive Order called for a full-scale national effort to enhance biomanufacturing, making products using programmed organisms at commercial scale.[6] In 2023, several government agencies, including the Departments of Defense, Energy and Agriculture, announced targets for using biology-made products to help bring the U.S. economy to net zero carbon emissions by 2050.

Some labs are in the clothing and housing industries. Some are in energy. Some are tool makers. Some sell custom-made DNA, and others, more outlandish ideas like self-assembling furniture. Meat without animals, seafood without fish, milk without cows, eggs without chickens, and sensors to assist your immune system—many are here or on the horizon. Some researchers sought to create life. Some won Nobel Prizes. Some were sued as tech giants fought upstart competitors. Other companies cloned beloved pets and prized polo ponies or racehorses. They are in a race to help solve some of the world's most pressing problems while potentially creating new ones.

Dystopia or utopia, what will it be? How might these discoveries provide energy sources or sustainable methods of cleaning our spills and waste? Are these products truly economical or environmentally friendly? This book explores those questions and the personalities of the people answering them. Its theme is that synthetic biology offers the potential for a change in our relation to the planet's resources, from one of exploitation to one of cooperative stewardship of biology's powers.

Part I introduces five breakthroughs in basic science—metabolic engineering in chapter 1, standardized cell parts in chapter 2, gene editing in chapter 3, directed evolution in chapter 4, and semisynthetic organisms in chapter 5. Part II traces synthetic biology's applications, in medicine in chapter 6, the environment in chapter 7, in food and clothing in chapter 8, in mining and the military in chapter 9, and in viruses as medicines in chapter 10. Part III retells the COVID-19 vaccine race as a case study of synthetic biology industrialization. Chapter 11 describes the creation of the vaccines, and chapter 12, their manufacturing. The ethics of altering life is covered in chapter 13, and its uses in space in chapter 14. We end with the main question: Do these new sources of medicines, food, energy, clothing, and shelter constitute a revolutionary science?

For four years, I traveled to conferences and labs around the country, interviewing more than 140 scientists, and attended business, lab, and Zoom meetings. I became captivated by the prospect of a new industrial revolution. As more products entered the market, I sought to understand how research creativity was confronting some of the biggest challenges of our time.

My goal is for readers to become caught up as well, with science's confusing, frustrating, exciting, and life-changing potential. This is a story of people in a community. I write this as a novelist might, with characters and conflicts, defeats and triumphs, teamwork and bitter disputes. Researchers pursued a big idea: to choreograph nature to help solve Earth's problems. In describing their work, I connect the science to the personalities who made it, embracing research while questioning the instrumentalism from which it springs.

Some of what follows features remembered conversations and reconstructed incidents. In portraying such recollected scenes, I have tried to interview everyone involved in the events and square my account with those in the published papers, online journals, and news stories. I am thankful to researchers who so patiently shared their time with me. Any mistakes are strictly my own.

These researchers are racing to harness life's power in new ways, changing what a scientist is and what science can do. These ideas and technologies are shaping our future. The question is, What will happen next?

I

BEGINNINGS

1

A GLASS OF ABSINTHE

A Malaria Medicine

I view microbes as little chemical factories.
—Jay Keasling, University of California, Berkeley

O
n Saturdays, he had to scrape out the manure in the pigsty. For his entire Nebraska high school career, Jay Keasling's hands smelled of manure and dirt. When he graduated, his dad sold the pigs. No one else would do the job. Years later, Keasling thought of his Berkeley lab as a sole proprietorship farm, where in 2006 he developed one of the world's first commercial synthetic biology successes, a medicine for malaria.

On the family farm if something broke, you fixed it yourself. Growing up, the lithe, wiry, self-reliant Keasling graduated as high school valedictorian and began learning about a new science where researchers moved genetic pathways from one organism to another. He read a book about the San Francisco company Genentech and its biomedical breakthrough in the 1970s and 1980s, its scientists figuring out how to engineer bacteria and yeast to make insulin more sustainably than by the traditional method of killing thousands of pigs and cows.

Keasling decided that was what he wanted to do with his life. "I wanted to engineer microbes to save the planet and make products people needed," he told me.[1] From his undergraduate days in chemistry and biology at the University of Nebraska to graduate work in chemical engineering at the University of Michigan, Keasling dreamt of getting to California and repeating Genentech's success.

At the time, only a few researchers were thinking of cells as factories. Biology could be a new energy source, the more visionary of them suggested, like steam was in the first industrial revolution, petroleum and electricity in the second, and nuclear power in the third.[2] Keasling and a few others envisioned a world of cellular parts, as standardized and interchangeable as computer and electrical parts.

But how to take the next step? Some researchers were writing bits of new genetic code, as Genentech researchers had done, using the four DNA bases, or building blocks: adenine (A), cytosine (C), thymine (T), and guanine (G). They did so with slow, painstaking applications of what is called recombinant DNA, squirting chemicals called inducers into Petri dishes to invite bacteria and yeast to take up snippets of the new genetic code, splicing the snippets almost like bits of recording tape into the genetic material of the organisms. These researchers developed techniques to make the process more precise. By the 2000s, scientists had edited the genes of organisms as diverse as microbes, soybeans, and sheep in the lab and created widely used ingredients of cheese, detergents, and ice cream in the process.

With gas prices rising in the 2000s, the government came calling. The Defense Advanced Research Projects Agency (DARPA), the Department of Energy, and the National Science Foundation commissioned projects from Keasling and others to engineer the metabolism of cells, optimizing their ability to make useful chemicals. Starting in Berkeley and then moving to Emeryville, California, Keasling came to direct a new interdisciplinary institute and won a $42 million Gates Foundation grant to brew a malaria medicine from yeast. The medicine came from the wormwood plant, or *Artemisia absinthium*, which is also used in the making of vermouth and absinthe. For years, his team edited gene pathways, moved them into yeast, and coaxed the buds to ferment the chemicals to make artemisinic acid, the key ingredient.[3] Keasling cofounded Amyris, a company funded in part by the federal government, to increase the medicine's production. Many Keasling lab graduates worked at Amyris and went on to found their own companies to create fuels, medicines, and chemicals.

They succeeded and failed and tried again in a story that featured brilliance, arrogance, hard work, wild parties, and fortunes made and lost. It was only the beginning.

LA BIOLOGIE SYNTHÉTIQUE: "LITTLE SHORT OF ASTONISHING"

The cultivation of desirable traits in living organisms had been going on since the dawn of farming and livestock breeding. In 1912, ninety-four years after Mary Shelley wrote her novel *Frankenstein*, the words *synthetic biology* were used in a book and in a science pamphlet's title, *La Biologie Synthétique*, published by Stéphane Leduc, a doctor and biophysicist at the École de Médecine in Nantes, France. Leduc theorized that life originated in chemistry. To pursue that idea, Leduc conducted experiments involving salt in solution being transformed into crystals. To critics, Leduc's experiments, as the physicist and science historian Evelyn Fox Keller initially described them, were "illuminating in proportion to what may now appear . . . as their absurdity."[4] Leduc was more likely about one hundred years ahead of his time, as Keller reversed herself in a 2016 essay, "Knowing as Making, Making as Knowing." Anticipating the rise of today's synthetic technoscience, Leduc's experiments on "the self-assembly of structures," Keller concluded, "are little short of astonishing."[5]

The use of fermentation in yeast and bacteria for making wine, beer, cheese, bread, and pickles had been going on for thousands of years. Fermentation entails the breakdown of a food, usually sugar, by yeast or bacteria, a process often giving off heat and effervescence. Microbes producing chemical ingredients by fermentation found their first industrial impact in the wars of the twentieth century. When during World War I a desperate civilian head of the Royal Navy, Winston Churchill, saw his country running out of cordite, the smokeless gunpowder used in launching the shells of its big naval guns, he extended an urgent call to the country's scientists. The University of Manchester chemist Chaim Weizmann responded that he could use natural bacteria to make cordite's required acetone. He had a small amount of the product in his lab. Britain needed tons. With Churchill's support, Weizmann increased production in giant fermentation tanks in England, Canada, and the United States, creating chemicals that helped win a war.

Something similar happened with the first antibiotic, penicillin, before World War II. Discovered by accident in a contaminated Petri dish, the medicine, which arises from tiny amounts of fungus in molding foods such as bread, suddenly had to be produced in vast quantities to save the lives of injured soldiers. The technology for making it in fermentation vats was perfected in a research facility in northern Illinois, using a bacterium from a moldy cantaloupe brought in by a Peoria housewife.[6]

As with cordite, the penicillin bacteria were natural, not engineered, but the techniques of large-scale fermentation would lay the groundwork for the synthetic biology industrialization to come.

Jay Keasling was born in 1964 in Harvard, Nebraska. Located on an agricultural plateau on the state's east side, Harvard was named by a railroad employee to follow neighboring Grafton in alphabetical order. When the sweet-faced Jay Keasling was eleven years old, his mother was recovering from cancer. Driving home from a doctor's appointment where she learned she was free of the disease, she waited at a stop sign. The corn was high, and visibility was poor. As she pulled out, an oncoming vehicle, driven by her cousin, collided with hers. Both died in the crash. "It was tough on Jay," his father, Max, told interviewers on an episode of the TV science show *Nova*, describing Keasling and his mother as "very close."

Stricken by her loss, Keasling was a driven young man who loved science and also kept a secret. "Being gay in small-town Nebraska is difficult," he told the same interviewers. "People who were, if there were any, were certainly not out, and so you had no examples at all." Not until he was a professor at Berkeley did Keasling come out to his family.

The story of Genentech offered Keasling a vision for his career. Herb Boyer, a biologist at the University of California, San Francisco, teamed up with the businessman Robert Swanson to apply Boyer's discovery that bacteria could be engineered to ferment the building blocks of insulin, which was in high demand because of a growing diabetes epidemic. Until 1980, all medical insulin had to be culled from pigs or cows, at huge cost and with much environmental damage. A single Indiana farm sacrificed 23,000 pigs a year to make one pound of the lifesaving hormone. Genentech researchers engineered bacteria by re-creating human insulin's genetic code in the lab and getting bacteria to take it up. Genentech scientists made the genes by splicing DNA snippets into plasmids, which are rings of genetic material. The new DNA hijacked the bacterial protein-making machinery to manufacture two insulin chains. Scientists combined them to form the insulin molecule.

By 1982, the FDA approved the Genentech insulin, and the company outfitted a San Francisco airfreight warehouse with steel fermentation vats to make it in bulk. A new industry, modifying microbes to make medicines, was born.[7]

Genentech's 1980s success was followed by that of another nearby biotechnology firm, Applied Molecular Genetics (Amgen) of Thousand Oaks, California, in the 1990s. From *Escherichia coli* and in yeast, Amgen made ingredients for an anemia treatment and an immune suppressor for chemotherapy patients. Other companies made sustainable clothing dyes using chemicals produced by *E. coli* bacteria, rather than using the toxic metals of conventional dye-making methods.

Then came cheese. Cheese was always produced using rennet enzymes from the stomachs of calves, lambs, and goats. The enzyme in rennet breaks down the solid particles in milk, making it curdle. After FDA approval in 1990, researchers inserted the rennet-making DNA into bacteria, fungi, and yeast to get them to make the enzyme more sustainably than by killing off calves, to the point where most cheese in America today is made with microbially manufactured rennet. Even the thickeners in chocolate cookies were the products of metabolic engineering.

But how should this emerging technology best be supported? DARPA took a keen interest in that question. Founded in 1958 in response to the Soviet Union's launch of the Sputnik satellite, the agency played a crucial role in advancing technical breakthroughs like weather satellites, the global positioning system

(GPS), and the protocols of the internet. As early as 2000, the agency funded synthetic biology research into manufacturing of medicines, bomb sensors, and industrial chemicals.[8] It also sought methods to engineer bacteria and yeast to make the hydrocarbons of biofuels.

With such goals in mind in 2002 and 2003, DARPA sponsored an eighteen-month study by academics including Keasling, business leaders, and policy makers to create a road map for a new science. The study, involving seven workshops, ended with a briefing of the DARPA director in 2003. This initial effort failed in its intent to create a funding program, but it did produce a three-tiered plan including the design and fabrication of genetic material via DNA synthesis, the coordination of labor by agreed-upon standards, and the management of biocomplexity using engineering principles.

The first synthetic biology event, Synthetic Biology 1.0, was held at MIT in 2004.[9] The ideas from that small, informal conference led to the establishment of two organizations: the Synthetic Biology Engineering Research Consortium (SynBERC), founded in 2006, and the Joint BioEnergy Institute (JBEI), funded by the Department of Energy and located in Emeryville, California, in 2007. The National Science Foundation helped fund SynBERC, a semiannual gathering of East and West Coast researchers to discuss their unpublished ideas and results. At their meetings, new and experienced scientists argued and theorized about the industrial possibilities of biology. Their meeting sites alternated between Berkeley and Boston, the other center of American synthetic biology research. Key people from Harvard, like George Church and Pamela Silver, and from MIT, like the computer engineer Tom Knight, joined with West Coast thinkers "to cement a new community," Keasling told me. They were not interested in mere genetics but rather in life's complete "hierarchical set of principles," said UCLA's Wendell Lim, the associate director of SynBERC. They wanted to engineer biology with the plug-and-play precision of computer programming. "When your hard drive dies," Keasling told the *New Yorker*, "you can go to the nearest computer store, buy a new one, and swap it out. . . . Why shouldn't we use biological parts in the same way?"[10] Lim wanted even more. "I wanted to learn the rules by which DNA encodes an elephant,"[11] he told me.

At the same time, a parallel series of initiatives at MIT, led by some of the same SynBERC thinkers such as Tom Knight, Drew Endy, and others, led to a more radical set of institutions. These arose from a pair of whimsical December-quarter courses in which students ranging from undergraduates to doctoral candidates were tasked with making *E. coli* cells blink like a light. The problem was that, much as in the early days of the electrical industry, different groups were using different, incompatible tools and parts. To bring standardization to biological parts, Tom Knight created a web repository called BioBricks, which made available mail-order kits to make DNA and proteins. Out of the December courses, the MIT researchers founded the International Genetically Engineered Machine (iGEM) student competition to make synthetic cells and, in so doing, add units to the BioBricks, innovations I cover more in chapter 2.

Meanwhile at Berkeley, the SynBERC conferees would meet in newly built Stanley Hall or at the old, white stucco Claremont Country Club. In Boston after their sessions at MIT or Harvard, many of the students went out clubbing, often led by Keasling. One California meeting lasted into the early morning, and the van driver disappeared to fall asleep in the lounge. Keasling drove the stranded faculty back to their hotel. "It was fantastic," MIT professor Kristala Prather recalled of the group. "This really interesting cross section of individuals with foundational training in different fields were trying to make biology into an industrial force."

JBEI's mission, in the wake of rising petroleum prices, was to create fuels from plants.[12] The U.S. government had supported efforts to produce the gasoline additive corn ethanol, made by fermenting corn, but ethanol was expensive, damaging to older machines, and criticized as being harmful to the environment. To confront the energy crisis, the researchers focused first on prairie switchgrass, an idea Keasling seized upon for revitalizing the American Midwest. "I wanted to make the Midwest the new Mideast," Keasling often said.[13] Hydrocarbons are produced in the wax in plant leaves. The yellow of sunflowers, the red of tomatoes, and the scents of flowers come from the combining of two hydrocarbons to make things called isoprenoid oils. Amazingly, those same oils were the basis of a plant-based malaria medicine and of fragrances, solvents, and biofuels.

In the years thereafter, however, gas prices dropped, and suddenly the urgency for plant or microbe-based biofuels slackened. One new idea was to turn bacteria and yeast into factories for products such as cosmetic ingredients, flavors, and fragrances, to name a few. But before all that, researchers made a malaria medicine.

A GLASS OF ABSINTHE

Along riverbanks in Bhutan, silvery, woody, small-stemmed wormwood plants grow next to fences and in abandoned lots. In its pure form, the bitter-tasting extract of the wormwood plant is a poisonous ingredient of absinthe, the drink made famous by writers and artists of 1920s Paris. However, during the Vietnam War, Chinese doctors treating drug-resistant

strains of malaria in soldiers discovered that extract of wormwood (*Artemisia absinthium*) contained a highly effective antimalarial compound, artemisinic acid, the key ingredient of a
drug called artemisinin.

In Keasling's lab one day in 2002, a graduate student showed
him an article about artemisinin. Keasling had never heard of
it, but the artemisinic acid it contained was made of the same
isoprenoid oils his lab was making for flavors and fragrances. The
wormwood plant was expensive to grow, and farmers did not
cultivate it consistently. If researchers could coax bacteria to create the key ingredient of artemisinin, they could save thousands
of lives. Keasling swerved the lab to try making artemisinic acid,
first in bacteria, then in yeast. "Having grown up on a farm, I was
not afraid to go in and monkey with something. That molecule
changed everything," Keasling told me.[14]

The wormwood plant makes artemisinic acid in its trichomes,
bags at the base of the leaf. The cells produce the molecule in oily
sacs, much as cannabis produces tetrahydrocannabinol (THC).
"We reasoned those cells must be rich in enzymes that produce
the artemisinin," Keasling told me from his book-lined JBEI
office. A postdoctoral fellow in Keasling's lab, Dae-Kyun Ro,
identified the enzyme that could produce artemisinic acid. "That
was the eureka moment," Keasling said.

In Africa at that time, every thirty seconds a child died of
malaria.[15] Caused by a microbe that infects mosquitoes, the
disease was becoming more resistant to treatments and was
spreading in the 2000s. Many countries could not afford to
buy the plant-derived artemisinin to combat it. On Keasling's
Berkeley team and then later at Amyris, up to fifty postdocs
and other researchers worked for a dozen years painstakingly
editing and transplanting metabolic pathways from yeast,
bacteria, and plants to transport the *Artemisia* gene pathway

into bacteria. In 2003 they had their first success, which was published in *Nature Biotechnology*. In 2006, they published improved results in *Nature*, this time using yeast as their factory.[16] In 2008, Amyris licensed its engineered yeast strain to the company Sanofi to launch the large-scale production of synthetic biology-made artemisinin. The process was described in a 2013 *Nature* article.[17]

Keasling and his team had built a microbial factory by inserting new genetic code into yeast. The lifesaving antimalarial drug could be produced so efficiently that a dose cost pennies instead of dollars. Approximately 500,000 lives could be saved each year. "Jay's genius was twofold," commented the Berkeley biologist Adam Arkin, "First was using synthetic biology to make an economically unattainable drug to fulfill a public good."[18] Second was in training and directing a group of "amazing people working together," recalled Eric Steen, an engineer and SynBERC member who later cofounded a synthetic biology industrial company called Lygos.

Elsewhere the new science was taking off. The team of Harvard researcher George Church was proposing to resurrect mammoths to help restore grasslands to cool the Earth, a quest I cover in chapter 7. Others were seeking ways to get bacteria and yeast to make the proteins of raw materials for clothing and food. Everyone wanted to make the next artemisinin.

Keasling had cofounded Amyris with three of his graduate students to industrialize the technique on a grand scale. Receiving the artemisinin first grant of $3.7 million, the youthful cofounder Neil Renninger noted that he had never seen that many zeroes on a check in his life. But the Gates Foundation required that artemisinin be sold at cost. That meant Amyris needed other profitable products. It decided to make biofuels.

"GO AHEAD AND TAKE RISKS"

Neil Renninger remembered how, when he was growing up in California, his father sent him on recycling errands in the 1970s. His mom was a schoolteacher, and Renninger excelled in math and science. He made a first splash as part of an MIT undergraduate group winning at blackjack card-counting in Las Vegas. He would stuff $100,000 in the lining of his hoodie, he told a *Fast Company* writer. "The biggest thing I learned . . . was to go ahead and take risks," he said, "because if you fail, you'll land on your feet."[19]

Renninger suggested that Amyris modify yeast to make the healthful plant oil called farnesene for use in fuels and cosmetics. The company ramped up with DNA-injecting robots nicknamed WALL-E and R2D2. The company's plans required the first-time modification of many genes in an early test of synthetic biology as a business investment.

Amyris experimented with a variety of strategies to move plant gene pathways into yeast, including the use of viruses as delivery vehicles. It was arduous work. The synthetic creation of plant hydrocarbons in industrial amounts was difficult. Lab successes stumbled if production was increased by factors of fifty or one hundred. Amyris turned to sugarcane, a source for biofuel in South American countries, as a more efficient raw material than corn. But sugarcane did not grow widely in the United States, so Amyris built an expensive state-of-the-art sugarcane fermentation plant in Brazil.

Still, the work was challenging. Moving individual genes between organisms capitalized on naturally occurring mechanisms like cell-to-cell transfer of genetic material. Scientists studied these mechanisms in simple systems such as bacteria and viruses. Now scientists were transferring genes into organisms as diverse as rice and racehorses. In its biofuel effort, the company felt it needed an experienced executive to manage bigger budgets, business strategy, and day-to-day decisions. It found a former London-office British Petroleum executive, a native of the Azores Islands named John Melo, who seemed to fit the need.

"A MAGICAL PLACE"

Growing up on a volcanic island near Portugal's coast, John Melo recalled the shouts of men whenever whales were spotted. His uncles and father would rush to the beach and jump into canoes to hunt the giant animals. Of the bloody process of cutting away the energy-rich blubber, he told a journalist, "I remember the smell vividly." Whale oil powered the lamps of the nineteenth century. Then petroleum replaced whale blubber as an energy source. He wondered whether Amyris could replace petroleum. "What really hooked me," Melo said of the start-up, was "the belief that you could do anything with biology."[20]

Melo expanded the company's production of biodiesel. Amyris negotiated agreements with two Brazilian city-bus systems and improved its enormous sugarcane fermentation facility to meet the demand. The plant near São Paulo required a dizzying climb up a metal scaffold to reach the lid of its giant fermentation tanks. From the top, workers looked over vast acres of sugarcane. To expand into production, Amyris perfected the use of yeast as the fermenting organism. Yeast was cheaper than bacteria and more acceptable to consumers, being more familiar from brewing beer and baking bread. Into the steel tanks flanking a sugarcane farm, workers poured Amyris's special engineered yeast strains, which in two weeks devoured as much as 1.2 million liters of rich cane syrup to produce the chemical farnesene for biodiesel fuel. The industry publication *Biofuels Digest* called them "eco-emirs."[21]

The challenge was the huge amount of fuel they needed to produce. In moving from the lab to the in-country plants, they increased production more than a hundredfold, exceeding the industrial chemist's general maxim not to increase the volume of a reaction by more than tenfold or twentyfold.[22] Problems beset the Brazilian fermentation tanks, including invading wild yeast, frothing cell walls, and even the different qualities of water between that of California and that of Brazil. There remained also a lingering Silicon Valley culture of entitlement. A bad sign was a Lake Tahoe company holiday retreat where the younger men partied late into the night. "I don't even want to tell you what they're into," Melo recalled to a journalist. Melo retreated to his hotel room.[23] There were no more holiday getaways.

Going to 10,000-liter aluminum tanks three stories high in Paraiso, Brazil, and from there starting to build a 200,000-liter facility an hour north of São Paulo, Amyris strove hard to meet stockholder demands. To add to the difficulty, the company had gone public in 2010, early in its development. The São Paolo and Rio de Janeiro government contracts to buy biodiesel for their bus systems at $7.80 a liter was a great price, but Amyris was still losing money. The stock price soared to $30 a share in 2011, but falling gas prices and the inability to scale up production dropped the stock to $1.50 a share a year later.

Then the price of natural artemisinin crashed. Despite the company's increased production, "if natural artemisinin is cheap, then there's no reason to fire up a fermenter," Keasling told SynBioWatch, one of the field's new watchdog groups. Sanofi prepared to sell its artemisinin factory. At Amyris, stockholder demands, technical difficulties, and fluctuations in oil prices took a toll. It was losing money. People were laid off. Some shareholders sued Amyris for some of its market claims. The company's leadership decided to focus instead on making ingredients for consumer products such as cosmetics and fragrances.

Renninger argued that the company's designer oil called farnesene could also be sold as a healthful skin and cosmetics base. If a little water was added to the tanks, Amyris could increase output some 5 to 10 percent. The company began selling to cosmetics producers and then developed its own brands. By 2017, the consumer products were bringing higher profits, and Amyris's lotions and fragrances were selling so well that the company launched two branded lines, Biossance Cosmetics for adults and Pipette for children. Amyris made the best money on its own products. A giant Buddha presided over its corporate lobby, and the lunchroom featured 1950s diner décor where workers exchanged a "candy shop of ideas," recalled Tim Gardner, then the director of research operations and programs.[24] By 2020, Amyris was selling more synthetic biology products than all other new companies combined. Despite its past missteps, its new emphasis on standardizing processes "calmed the noise," Gardner told me.

Other companies were modifying microbes to make agricultural products, such as POET in South Dakota, ADM in Illinois and Cargill in Minnesota. In San Diego the synthetic biology company now called Geno made a variety of ingredients for household products. But in the working-class town of Emeryville, California, smaller start-ups were springing up to put SynBERC ideas into wider practice. They made medicines, clothing materials, and industrial products. In the late summer of 2019, as money first came in, I made my first visit to the town that was a center of a new industry.

BRAVE NEW WORLD

As an August morning fog was clearing in Emeryville, Keasling showed me around JBEI's steel-and-glass-clad institute. Pointing to the low clouds hugging the Berkeley hills, he remarked, "The Bay Area is a magical place." He greeted young people who stopped in the hall. "I'm not interested in mammoths," he added, taking a swipe at George Church's quest to resurrect the extinct giant. "I am interested in practical things to save the environment."

Still trim but now balding, the energetic Keasling wore Elton John–style large black eyeglasses, black pants, an open-necked shirt, and cream-colored cowboy boots. As we walked the halls, greeting more lab workers, he explained to me how the lab equipment was shared to enhance collaboration. "There's no principal investigators, no hierarchy," Keasling said. "JBEI's a completely socialist organization." Motioning to a neighboring room, he said "Here's where we made artemisinin." "Here are the machines where we synthesize DNA. We're not as slick as what you'd see in commercial labs, but many of them got their start using our facilities."[25] We passed into a plant-growing room with giant *Arabidopsis* (a mustard plant with a small genome) and grain sorghum, or switchgrass, grown in Texas, Missouri, Kansas, and the Dakotas.

"This is entrepreneur heaven. We are surrounded by start-ups we fostered," Keasling told me. "Amyris is in the same building, Zymergen is across the street, Bolt Threads and Demetrix, another company I helped start, are down the block." Lygos, the automobile finishing and industrial products company, was several blocks away.

Amyris employed at least four of Keasling's former postdocs, and Lygos was run by two of his former graduate students, Eric

Steen and Jeff Dietrich.[26] Christina Smolke, CEO of the medical products manufacturer Antheia, came from his lab. The bioengineer Kristala Jones Prather at MIT is another graduate of his Berkeley lab and cofounder of a metabolic engineering company based in Massachusetts. "I'm not taking credit," Keasling told me. "One of the most important things about my lab is not all the ideas had to come from me."

Demetrix was making rare cannabinoids from yeast products to treat chronic pain. Bolt Threads was making clothing from yeast secretions and later from mushroom roots, science I cover in chapters 4 and 8. A third, Zymergen, built a facility and employed 800 people.[27] Founded by a couple of former Amyris scientists, it was promoting a microbe-made product to be used in foldable device screens. That would not work out.

The industrial lab Lygos made a key automotive finish ingredient, malonic acid, from corn.[28] "We founded Lygos with the mission of taking raw materials like sugar derived from Midwest corn and producing a product that ultimately helps the Midwest," the Indiana-raised Eric Steen told me.[29] The idea was "to help bring chemical manufacturing back to the U.S. And do it in a way that's better for the world. We're not using toxic processes. We're not going to produce products that cause cancer." One of Lygos's investors was the Department of Energy.[30]

For its part, Amyris fought its way back. Employing 600 people, it sold synthetic biology products that included ingredients for diesel and jet fuels, polymers, surfactants, coatings, adhesives, and solvents, as well as cosmetics products, baby wipes, and shampoo. It had developed an understanding of the economics of synthetic biology products and partnered with cosmetics manufacturer Sephora. One key pharmaceutical oil it made was squalene, normally found in shark liver and plants such as olives, and in human skin. Squalene must be hydrogenated into the product called squalane, which is an ingredient in vaccine enhancers. When the COVID-19 pandemic hit in 2020, squalene would come into high demand.[31] By 2022 the company opened a new biomanufacturing facility in Barra Bonita, Brazil, featuring five smaller fermentation mini-factories to speed up production.

Keasling was still all-in to make biofuels, but not from corn, which was criticized for being no more sustainable than petroleum. "You could use the rest of the corn plant, or perennials" that do not need irrigation and fertilizer, Keasling said. "Sugarcane is great. Farnesene was such a great fuel, but it was too expensive. Oil already has its $100 trillion infrastructure of drilling, transportation, refineries, and sales, where biofuels must build new infrastructure piece by piece."[32] Another company Keasling co-founded, Zero Acre Farms, produced a sustainable cooking oil in 2022.

Synthetic biology as a business was a roller coaster. What it needed was a philosophy or unifying theme. This would come from across the country.

2

A RADICAL PHILOSOPHY

Biology is the world's greatest chemist.

—Pamela Silver

H arvard's Pamela Silver grew up in the small California town of Atherton where her guitar teacher, from the local music store, was the Grateful Dead's Jerry Garcia.[1] Silver rode horses and enjoyed doing experiments at an all-girls prep school, eventually discovering a lifelong love of math, science, and sailing. But her parents' marriage was falling apart, and she was left on her own much of the time, learning an independence that helped her craft a unique career as a pioneering synthetic biologist.

She was raised in what would later become the center of Silicon Valley, but Atherton in the 1960s was mainly a small California town. Her father had studied psychology at Harvard Medical School, where he never felt comfortable, choosing to escape to the West Coast to practice. It was a heady time for science in their town, with Steve Jobs and Steve Wozniak in a garage and rock and roll in the clubs.[2] Her father brought a dissected cat to Silver's first-grade class and held a weekly poker game with Nobel-winning chemist Linus Pauling, in whose clinic Pamela had an early job. On weekends, she and friends escaped to San Francisco to see local bands like the Grateful Dead. At the University of California, Santa Cruz, she raced a sailboat.

Like her father, Silver made it to Harvard, in her case for postdoctoral research in chemistry. Unlike others at the school, Silver had no master plan for her future scientific career, but when she took a renowned Woods Hole, Massachusetts, summer course in genetics, she was smitten. Famed geneticist Barbara McClintock guest lectured, and Silver "loved the camaraderie, staying up all night and discovering something new about life,"[3] she told me.

From there she joined the research group of Harvard molecular biologist Mark Ptashne, who would sometimes bring his violin to campus and play. There she studied the ways proteins in the cell moved in and out of the nucleus, trying to apply her research to cancer therapies. "She was not afraid to fail," one of her later doctoral students, Christina Agapakis, remarked about Silver's boldness. Silver pursued her idea that certain amino acids targeted proteins to the nucleus.[4] "It was a wacky idea," Silver recalled, "but I was thinking . . . 'well, I'm doing this thing and it might fail and I'll just have to find a job doing something else,' which is itself kind of scary."[5]

She went on to a distinguished career as a biology professor, first at Princeton and then at Harvard, where she became friends with thinkers trying to apply engineering concepts to biology, including the computer designer Tom Knight and a young engineer named Drew Endy. After he was hired as a professor at MIT, Endy slept in her basement because he had no other place to stay. It was a moment, recalled Caltech's Michael Elowitz to me, "when we physicists and engineers thought we could solve biology." Biologists told them life did not work that way.

Silver and MIT researchers like Knight, Endy, and Randy Rettberg helped to found a new field and named it synthetic biology. This group held some of the field's first conferences, founded its international student competition, began trying to standardize its parts (proteins and genes), and promoted its new ethos of freely sharing ideas. If Keasling and artemisinin in California marked the first triumph of metabolic engineering, this group of friends and rivals, idealists and profit makers, sought to make synthetic biology a radicalizing discipline, transcending scientific hierarchies. They talked to artists, designers, do-it-yourself biologists, and teenaged scientists because they felt the idea was bigger than science. It demanded more of its researchers than the usual pursuit of papers and prizes. "Pursuing such work will require the widespread societal acceptance of responsibility," Endy wrote, "for the direct manipulation of genetic information."[6] Synthetic biology could be a communal effort to rewrite genomes for a more sustainable future. "We need to feed the world," Silver told me, "and synthetic biology is the way to do that."

At the very beginning, though, two pairs of outsiders working independently of each other achieved an obscure breakthrough. When their papers appeared simultaneously in 2000, one reviewer called the results "pinheaded." But approved or not, their breakthrough detonated science by showing that cells could be programmed like computers.

CIRCUITS

When Los Angeles native Michael Elowitz was at Hamilton High School in the late 1980s, he and his brother would hang out at a diner with their social studies teacher. They were creating a computer version of their teacher's world domination game.

Elowitz attended the humanities magnet within the high school, which was "not science-focused, but really creative and funny and weird."

Elowitz obtained a PhD in physics at Princeton, where he joined a group of biologically minded physicists in the lab of Polish-born researcher Stanislas Leibler. Liebler created an extraordinary scientific environment dedicated to unlocking the operations of life. Their physicists' arrogance was that the prevailing way of doing things in molecular biology was mostly "cataloging molecular interactions," Elowitz noted, but leaving "deep questions unaddressed." The idea was to do for biology what Apple was doing for personal computing, to make it easier to program. "In biology papers at the end, there was often a diagram," Elowitz noted. It would feature arrows offering what the researcher thought was going on, but "it was unclear if the arrows really explained cell behavior," Elowitz said. "The only way to answer that was to build the gene circuit and see if does what you think it does inside a cell."[7]

A genetic circuit is an assembly of biological parts that enables cells to perform a function. Elowitz and Leibler started with the microbiology workhorse *E. coli*, using the same gene first activated and deactivated by Nobel-winning French biologist Jacques Monod in the 1950s. But Monod's was a binary switch that had to be continually set to "on" to work. What they were trying to make was a switch that went on and off in a periodic cycle, an oscillator like a biological Foucault pendulum. "In physics, oscillators are everywhere," Elowitz explained. "Oscillators make the waves on a beach or in an electromagnetic field, or life's repeating circadian clocks" that tell animals when to wake, sleep, eat, or migrate thousands of miles.

Unknown to Elowitz, Boston University's Jim Collins was trying much the same thing, to turn a gene off and on like an electronic switch. Collins sought to build a genetic circuit in an *E. coli* bacterium using the same well-characterized genes Elowitz was using. His doctoral student Tim Gardner, who went on to become Amyris's computational biology director, started by trying to mimic Luke Skywalker's prosthetic hand from *The Empire Strikes Back*. "A mechanical hand was too difficult, so I turned to making genes to grow one,"[8] Gardner recalled to me. The first challenge was turning on a gene and keeping it on. Thinking about electronic switches while on a bus from Boston to New York, Gardner hit on a solution. "I realized I could use two genes and two promoters (sequences that turn a gene on and off). I was dumb enough to think I could do it."

While Gardner struggled, Collins sought support from the Office of Naval Research. "I reached out," Collins told me from his blue-walled home office. "The program officer said, I don't have any money. If I did, I absolutely wouldn't give it to this." Collins called him monthly for the next nine months. "Finally, he gave us $500,000 so I would stop calling." But there was a catch. His student Gardner had to present the data to the Office of Naval Research in front of science luminaries, even before their programmable gene switch was finished. As Gardner was speaking, the Nobel laureate Sydney Brenner raced up to the blackboard to show why their design was all wrong. "I should have been ashamed, but to me it was cool, like a scene from *The Making of the Atomic Bomb*." Gardner smiled, thinking of his favorite classic popular book about the atomic race.

Meanwhile at Princeton, Elowitz and Leibler were using the same genes to make their oscillator turn on and off in a repeating[9] "rock, paper, scissors" sequence, Elowitz told me of their attempt to build a living clock in a cell. "I had read Richard Dawkins's *The Blind Watchmaker* and was entranced by the metaphor of life's clocks," he recalled. To judge whether his *E. coli* timer was working, he tagged his system with the gene for green fluorescent protein. He was working long hours (Elowitz cowrote a song with a Liebler postdoc, "Sunday at the Lab.") It was wintertime, and he slept in the lab at the snowy Princeton campus to keep track of the fast-reproducing bacteria. The bacteria manufactured copies of his switch every time they divided. Because the edited genes were inherited, he watched the *E. coli* offspring.

Around three o'clock one morning, Elowitz stumbled from his cot to the microscope in the dark, cold room. He could not believe his eyes. Under the lens, the bacteria glowed green, on and off, pulsing in a regular, ghostly oscillation, "every 150 minutes," he told me. It was an eerie living light. The period was so long he had to record it and watch a sped-up replay. The result, in Elowitz's words, "was like a hydrogen atom for biology." As hydrogen was a simple atom that made complex materials like water, "so too do we find these simple circuits reappear in more complex biological systems. They provide design principles to understand those more complex systems." Two different teams programmed the same bacterium to turn a gene off and on like a binary switch, using the same genes but each in a totally different process. It was "as if our antennae were picking up the same idea out of the ether," Gardner told me.[10]

In 2000, the two papers were published back-to-back in *Nature*.[11] The fact you could build a synthetic circuit in a cell that turned on and off opened a new world of programming cells to sense and compute inputs. "The complex and practical design of synthetic gene networks," wrote Gardner and Collins, "is a practical and achievable goal." Most science historians date the beginning of synthetic biology to this breakthrough.[12]

The breakthrough opened eyes for many young researchers. To Stanford student Mo Khalil, who would become a PhD candidate with Collins, it was a turning point for two reasons.

"First, the fact that you could build a circuit to alter cell function illustrated an inherent modularity in biology. Second, you could alter function with components that were not really evolved to do that," he told me. "Both featured beautiful mathematical models simple enough to write, but complicated enough to capture the nonlinear cooperativity of cell components. Applying those concepts of engineering to biology was inspiring to people like me."[13]

The community grew. The SynBio 1.0 conference at MIT was followed by SynBio 2.0 in Berkeley in 2006. SynBio 3.0 took place in Zurich in 2007, and SynBio 4.0 was held in the sleek, towering conference center at Hong Kong University in 2008. By that time, it was a huge conference, observed researcher, author, and entrepreneur Rob Carlson. "What a difference a few years make," he wrote in his blog. The ideas were catching fire.

Another entity took notice. DARPA now funded several synthetic biology projects, as did other federal agencies willing to invest money to devise new strategies to enhance soldiers or to maintain and recycle equipment. Genetic circuits were only the beginning.

More was coming from an engineer with a poetic sense.

A "SENSE OF THE SACRED"

Growing up in Pennsylvania, Drew Endy liked walking in the woods and playing with Lincoln Logs, Legos, and Erector sets. He almost failed high school biology and had a summer job building bridges for Amtrak. In the mid-1990s as a doctoral student at Dartmouth, he modeled a well-known bacterial virus, called the T7 phage, on a computer. He sought to understand its function, but his computer program did not accurately predict the virus' behavior. Rather than redesign his computer version, at MIT he asked graduate students to build a surrogate version of the natural virus, which they succeeded in doing in 2005.[14] "I was infected by that virus," he would say in lectures.[15] At MIT, Endy began meeting with computer programmer Tom Knight, who was also fascinated by life's computational power. Together, the two sought to create a philosophical framework for a new science. Three years after, MIT's Biological Engineering Department was created, Endy was hired as a tenure-track faculty member. To him, this was more than merely research. Later, teaching at Stanford, he opened and closed a TED Talk by extolling the beauty of a pine cone that came from his yard, insisting in an interview that studying synthetic biology required a "sense of the sacred."[16]

Endy worked with Tom Knight, the well-liked professor who grew up a 45-minute drive north of Boston and had attended MIT as an undergraduate and graduate student, even taking courses when he was in high school, and was a co-engineer of ARPANET, the precursor to the internet. Knight was a widely respected computer programming pioneer. But by the early 2000s, he was becoming more excited by the computational power and design capabilities of biology. "It's time for a rewrite," he famously repeated, of the DNA code of life. Editing nature like a computer seemed "like the obvious next thing to do," he told conference audiences.[17]

In East Cambridge and Boston, Knight and Endy began collaborating with others, including MIT's Randy Rettberg and Harvard's Pamela Silver. They formed a Synthetic Biology Working Group in part to agree on methods for the field. To mature into a science field, Silver said, "you needed standardization. . . . It was a bunch of Just-So stories that anyone with a DNA sequencer might announce in the *New York Times*."[18] To address that issue,

Knight and Rettberg established the BioBricks repository to offer free, standardized *E. coli* plasmids (circles of DNA used in research) and access to data on synthetic biology discoveries. The repository shared promoters and primers, snippets of nucleic acid that induce the creation of proteins or DNA, also for free. Randy Rettberg created a Registry of Standard Biological Parts on the basis of their work.

Together they conceived the International Genetically Engineered Machine (iGEM) student biology competition. iGEM inspired a fun, risk-taking philosophy for the emerging field. Starting in 2004, student teams gathered from around the nation and in 2005 from around the world to compete for prizes, dressing in matching T-shirts and making humorous videos to explain their synthetic biology experiments. For some, coming to the United States from developing countries, it was the first encounter with other young people like themselves—young women from Nigeria, a high school class from Belgium, teams from China and India—all trying to make biology solve world problems. By 2008, a thousand students competed. By 2012 the number doubled, and then doubled again. The ideas they proposed ran the gamut in quality, but the spirit of camaraderie and creativity set an exuberant tone. By 2022, iGEM numbered 60,000 graduates.[19]

The BioBricks repository joined a clearinghouse founded by MIT graduate students for exchanging protocols or methods, called OpenWetWare (the term for living biology experiments), with three goals: to lower the technical barriers slowing down the publication of synthetic biology discoveries; to build a community that would practice the open sharing of information; and to create new reward incentives in research to foster faster sharing of discoveries. Labs had to apply for membership and could make use of Open Wetware's tools, again for free.

The free-use idea also attracted a community of do-it-your-selfers who created a Facebook page, DIYBio, which covered issues like the best fermenting organisms to make kombucha. The DIY movement came partly out of the home fermenters and bakers of the slow food movement, and partly out of the Occupy Wall Street protests of 2011. It rejected the commercialization of molecular biology. By the mid-2010s, cities from Copenhagen to New York featured "community labs" where people could take classes and do free genetic experiments on their own. Some of them competed in iGEM.

Year after year, still larger groups of students joined in, and some of their ideas became commercialized. The staff of Bluepha, a Beijing-based biotechnology company, included members of a 2010 iGEM team. The company engineered microorganisms to make chemicals for bioplastics. The project of an iGEM student team with members from Paris led to a company named Eligo Bioscience, which partners with Glaxo-SmithKline. A company called Opentrons competed at iGEM to build a cheap, easy-to-use lab robot. "I want to change how we make things," proclaimed Endy, including sending DNA over the internet, sharing how biology "puts atoms in a string with atomic precision." The toolmaking company Bench-ling works with many iGEM teams as an official partner, and Canada-based cannabinoid producer Hyasynth also originated at iGEM, as did the celiac disease therapeutic outfit PvP Bio-logics, which originated from a 2008 University of Washington team led by Ingrid Swanson, who later sold her company for half a billion dollars.

In the 2020 competition on Zoom, the finalist teams from France, China, and Lithuania offered ideas ranging from disease-prevention in fish, to providing astronauts with nutrients, to decreasing pesticide residues in tea.[20] The 2021 competitors included a Hangzhou, China, high school team that engineered a plastic-eating microbe.[21]

After COVID-19, the competition also encouraged teams to create biodefenses. A 2021 Nairobi community lab, for instance, won an award for its microbial water quality sensor that could be used to thwart terrorist contamination. The projects advanced tempting ideas like modifying gut bacteria to look for cancer tumors or houseplants to produce a glow sufficient to illuminate a room.

During this time, several synthetic biology companies were founded. But exactly what products would sell most readily, and at what price, and to whom? From the very first iGEM, a big idea for a new kind of biological assembly-line company was brewing.

"LIVING CHEAPLY"

By 2008, Tom Knight was wearying of academia. When four MIT graduate students approached him with the idea of starting a synthetic biology company, Knight asked if he could join. Two students were from Drew Endy's lab, Jason Kelly and Barry Canton, and two were from his lab, Austin Che and Reshma Shetty.[22] Biological engineer and *Jurassic Park* aficionado Jason Kelly from Florida, inspired by an Endy lecture, had chased Endy down a hallway to ask if he could work with him for his doctorate. Kelly, along with Shetty (a smart engineer raised in Salt Lake City), Canton (a dark-haired Irishman), Che (a Stanford-educated contrarian), and Knight had advised a first-year iGEM team that created sweet-smelling *E. coli*. They knew each other well, and Canton was dating Shetty. For his part, Kelly had sought to be a genetic engineer since "I was a child," he told me.[23]

They decided they wanted to form a company but did not know how. They twice entered the MIT $100K Entrepreneurship Competition without advancing past the first round. They applied for National Institutes of Health (NIH) grants but were rejected. They gathered to discuss their future in a conference room in architect Frank Gehry's stunning, cantilevered Stata Center for Computer Design. What was it they could sell? The economy in those years was tanking. The usual optimist Kelly wondered whether they should quit. Outside in the neighboring streets, signs glowed from the façades of biotech successes, along with hallowed research centers like the Broad Institute. According to Shetty, they had "no technology, no space, no marketing plan." Investors had "laughed us out of the room."[24]

They decided not to make products themselves. They were expert engineers, and what they knew was how to build tools. Why not build a factory in which others could design products? In the meantime, they would self-finance. "We could bootstrap," said Shetty, "do consulting gigs while we figured out the business."[25] They grabbed discarded equipment from biotech companies going bankrupt. They found cheap space initially in a U-Haul storage locker.

Knight invested $150,000 of his own money and, as they won a National Science Foundation grant, they grew to about twenty people over the following few years. "A depression is an excellent time to start a company," Knight recalled to me at the 2019 Ginkgo-sponsored Ferment conference reception overlooking Boston Harbor. Knight proposed the name Ginkgo, for the prehistoric Chinese tree with no relatives on the tree of life, and they called their labs "biofoundries," for customers to design and make microorganisms they wanted. They would make biology easier to engineer.

Moving into the spacious new Boston Design Center on the harbor, they grew from 4,000 square feet to 50,000, and then added 50,000 more. They won Department of Defense research support and a $5 million Advanced Research Projects Agency grant that enabled them to build their first biofoundries, Bioworks1 and 2. At first their goals were modest. The biofoundries could be used for "things like engineering yeast to feature new rose aromas," Shetty told *Fierce Pharma*, a biotech industry newsletter.[26] At that 2019 Ginkgo Ferment conference, the bar featured cocktails like The Thai Breeze, made with vodka and "Natural Thai Breeze Flavored Syrup" Prototype APF0016418.[27] They paid themselves modestly to start. A key to their survival, Shetty said, was "living cheaply." That would change.

A BIONIC LEAF

In 2011, Harvard's Daniel Nocera created an artificial leaf to generate energy from sunlight. The cell-phone-sized device split water into its components of hydrogen and oxygen, much like leaves do; his idea was to burn the hydrogen for energy. But the system could not store energy. To do that, Nocera, a chemist, decided he and his students needed to learn synthetic biology.

In 2016, Nocera added hydrogen-eating bacteria that his students had learned to engineer in Pamela Silver's lab at Harvard Medical School. They had trekked from the main Harvard campus to the medical school across the Charles River to figure out how to use the bacteria to make a "bionic leaf." "I wanted literally to make liquid fuel," Nocera told me from his home, wearing a bright tie-dyed turquoise and yellow shirt. "I said look, bacteria, your only food source is hydrogen. That's going to give you energy. I want you to breathe in carbon dioxide and grow."[28] The resulting biomass would be used to produce a liquid cousin to ethanol for fuel.

In 2019, a doctoral student in Nocera's lab pointed out that some hydrogen-eating bacteria naturally produce ammonia and phosphate, both powerful fertilizers. The product could be a yellowish bacteria-laden liquid, poured onto fields, to which one added a compound to get the bacteria to produce ammonia. "You just need sunlight, air, and water," Nocera said, "and you can do it in your backyards." Again, Nocera's students worked with those of the Silver lab to develop bionic leaves into fertilizer factories.[29]

Other labs working on models of different kinds of bionic leaves included the California Institute of Technology, the University of Waterloo in Canada, and Cambridge University in England. Possible uses for the bionic leaf encompassed water systems needing low voltages, fertilizer production, carbon removal systems, and sustainable building façades. From California, to Boston, to London, Paris, and Beijing, the nascent technological ideas were taking hold. Researchers' next step was to persuade governments to fund development and distribution. For that, you needed a philosophy.

A RADICAL PHILOSOPHY

Several thinkers tried to develop a philosophical underpinning of the new science, especially as genetic tinkering in animals was getting closer and closer to human gene editing. "What issues about the self are raised by the creation of new biological entities," asked University of Chicago scholar Laurie Zoloth, of a science "with the promise of changing the world . . . for the improvement of human beings?"[30] Such a philosophical under-pinning could be based on the environmental movement or on the much older traditions of moral philosophy. The question was, What motives should drive synthetic biology and for whose gain? There had been several failures, such as the attempt to make biofuel from algae, which had attracted hundreds of millions of dollars from 2005 to 2017 before stalling, only to be revived by ExxonMobil with a Department of Energy grant. Synthetic biology mechanisms remained expensive and crash prone. How would you sell them to policy makers and the public?

Supporters faced new challenges after the anthrax and similar pathogen scares. Fears of garage bioterror grew even as both private and public funding increased. As Endy told a TED audience, "We were caught . . . in half-pipe of doom" conversations.[31] The philosophical backdrop to related fields like botany, idealistic as it was, lay partly in European imperialism, researching nature for the uses of man while potentially depleting natural resources. Some synthetic biologists sought to reverse that ethos. One of Ginkgo's early projects was to resurrect the putative scent of a Maui flower that American colonial development caused to go extinct.

Yet start-ups required, and attracted, huge amounts of money. Raising money meant that profit became the measure of success. For years, Amyris's former CTO Neil Renninger complained the company's problems stemmed from overpromising a speedy return on investment. The ethic of BioBricks would not work with a start-up company's need to patent discoveries. Investment capital drove research toward niche-market drugs that could be priced higher than, say, vaccines or antibiotics. For that reason, the idea of synthetic biology–made vaccines was downplayed at first.

Endy, Silver, Knight, and others insisted that many of the discoveries of synthetic biology be available free, like the Linux operating system. Endy explained his idea to *Discover* magazine, citing Wikipedia as a model of a joint effort to share genetic information. Shetty once asked Sophia Roosth, the historian of the early science, "Where can you publish just ideas?" It would be easy to criticize such idealism after fortunes later poured into such companies. By 2021, for instance, the Queens-based robot company Opentrons, cofounded by Occupy Wall Street protester Will Canine, was worth more than a billion dollars. "There are many avenues to your goal," he told me.

The idealism had an effect. iGEM inspired tens of thousands of young people, many of whom are working on technologies to address the climate and energy crises. All of their projects must include a statement on human impact. When two early human embryonic gene editing meetings were closed to the public, Endy and Zoloth published a letter of complaint in a 2016 issue of *Cosmos*. The next meeting was open.

As accelerating discoveries outstripped ethics rules meant to guide them, the science became centered in the main existential questions of our time, those of access, shared benefits, and fair pricing.

Two quieter, earlier synthetic biology businesses had already found success while promoting idealistic goals.

IN THE CORNFIELDS

Among the sprawling athletic fields and noisy exurban malls of Coralville, Iowa, Integrated DNA Technologies (IDT) produces high-quality synthetic DNA for products, medicines, and research. Founded in 1987 by Prof. Joseph Walder of the

University of Iowa, IDT was one of the world's first providers of oligonucleotides, or oligos, which are short DNA or RNA molecules. Oligos are used for molecular biology applications, ranging from genome editing to digital data storage. Yes, DNA was being tested as a repository for digital data, 1 gram of which could potentially hold 1 trillion gigabytes, a science quest I cover in chapter 5.

IDT describes itself as "scientists working for scientists."[32] The soft-spoken Walder attributes his interest in helping others to his Orthodox Jewish faith. His deep-pooled eyes reflect concern as he speaks of scientists' responsibility to innovate for the common good. When Walder earned his PhD from Northwestern University in 1978, the beginnings of nucleic acid synthesis and genetic sequencing were only ten years old. Realizing that it would take "an army of postdocs" to synthesize a minimal number of DNA fragments, Walder opened a lab focused on producing oligos at an industrial scale so that "biologists could concentrate on biology." At IDT, he and his team reduced the cost of the product through efficiencies and improved technology, making experiments requiring hundreds of thousands of oligos feasible for fellow scientists.

From the prairie to plants in Belgium and Singapore, IDT manufactured and shipped DNA across the world quickly and efficiently. IDT's vice president of global operations, Christine Boge, told reporters, "We are still a smaller company, so we are able to be more agile." In its thirty-six years, IDT has served more than 130,000 life-science researchers in more than 100 countries, revolutionizing the development of agriculture, medical diagnostics, pharmaceuticals, and synthetic biology.

A similar success of synthetic biology took place in Woburn, Massachusetts, at another company started in the early years. Sample6, cofounded by Boston University's Jim Collins, engineered phage viruses to detect bacterial contamination in food produc-tion. With some forty customers by 2016, including Unilever and the owners of Ben and Jerry's ice cream, Sample6 operated on the premise that food safety is "still in the age of the type-writer," according to Acre Venture Partners investor Sam Kass.[33] The company had "the potential to create a paradigm shift," Kass said, "from days-long wait times . . . to in-shift testing with immediate remediation following."[34]

IDT and Sample6 operated mostly outside of media cover-age, but big pharmaceutical companies like Bayer and agricul-tural manufacturers like Cargill paid attention. As competition and stakes increased, some thinkers focused on the larger mes-sage for a new industry that was coming into its own.

"A WHOLE NEW FOOTING"

Back in Massachusetts, Pamela Silver's students were engineer-ing human gut microbes to detect disease. Another important Silver project was a photosynthetic clothing collaboration with MIT-based artist and media studies professor Neri Oxman. Others of her students took on the roles of philosophers of the new science.

Silver's former graduate student Christina Agapakis, who became creative director at Ginkgo Bioworks, wrote frequently about the ethics of synthetic biology as the company grew. Ginkgo signed a first deal in the flavor and fragrance industry. A big early turning point had come in 2014 with the first biotech investment by the Y Combinator fund, a clearinghouse that had invested in giants like Airbnb. In 2015, another round of Ginkgo funding came from Viking Global Investors. "That was a larger injection of capital than we'd ever had," Shetty told me. "It put us on a whole new footing."

They expanded their facilities in the huge, retooled former military warehouse on Boston Harbor called the Boston Design Center. Bioworks3 opened there in December 2017. The company decorated the glass walls with dinosaur stencils, and employees brought their pets to work, where a parenting station was available to them. Borrowing terminology from *Star Wars*, they called their lab workers "padawans" and the scientists who supervised them "Jedi." Reshma Shetty featured the *Jurassic Park* logo in presentations. An agreement with German aspirin giant Bayer in 2017 put them further into the front of the business stage.

Many of the new forays were building upon a gene editing breakthrough called CRISPR for clustered regularly interspaced short palindromic repeats. Gene editing was essential to synthetic biology, and CRISPR was a fast, cheap, and effective way to do it. This biggest breakthrough in microbiology since the discovery of DNA came from an unlikely place.

3

PANDORA'S BOX

The Triumph and Temptation of Gene Editing

We cracked open the door. Now the crack gets wider.
—Haydar Frangoul, pediatric oncologist, TriStar Medical Group

Growing up in Mississippi, Victoria Gray could not play like other children. A gene mutation caused her red blood cells to degrade and clog her bloodstream. Not enough oxygen was delivered to vital organs, leaving her prone to infection. Like the 100,000 other sufferers of sickle cell anemia in the United States, she lived a life riddled by hospital visits for blood transfusions. As she reached adulthood, Gray worried about surviving past middle age.[1] Then she heard about a study of a gene editing therapy for which researchers were seeking subjects. The thirty-four-year-old mother of four rushed to volunteer.[2]

The saga of the gene editing tool called CRISPR is one of advancement from pure science to the clinic, and from big business to the courtroom, all in a short time. To edit a gene, scientists cut damaged DNA and splice a repair that is incorporated into the cell. The procedure won a 2020 Nobel Prize for Berkeley's Jennifer Doudna and the Max Planck Society's Emmanuelle

Charpentier, vaulting synthetic biology and biotechnology into a new level of public awareness. The Nobel Committee noted that "these genetic scissors have taken the life sciences into a new epoch."[3] In his book on Doudna, author Walter Isaacson termed it "the most important discovery since DNA." It offered to save lives like those of Victoria Gray.

It all began with a seemingly obscure discovery that catapulted an age-old dream, curing heritable diseases, to the verge of coming true. It paved the way for the synthetic biology revolution by offering a simple, inexpensive, and universal platform to program life. It also opened a Pandora's box of troubles.

In a Nashville clinic, Victoria Gray received an infusion of her own, altered bone marrow cells, each cell featuring an edit done with a microscopic genetic scissors.[4] The editing technique, almost as simple as cut-and-paste in a Word document, can be used to edit genes in almost all plants and animals. Many CRISPR efforts, like the editing of Gray's bone marrow cells, were heralded. But then the talk turned to prenatal manipulation of the genetics of human embryos.

DNA consists of two strands of four nucleotides twisting around each other, as we recall from chapter 1: adenine (A), thymine (T), cytosine (C), and guanine (G). They are held together by a simple rule: "A" matches with "T," and "C" matches with "G." Previous DNA editing mechanisms were fashioned to edit one gene at a time, or sequence of A, C, T, and G that encodes a product, a highly laborious task considering that sometimes hundreds or even thousands of genes are involved in most illnesses. Each time researchers wanted to edit a new gene, they had to craft a new technique. CRISPR solved that problem by offering a single platform to tweak any gene in any living thing at any time. It advanced rapidly from the original version, which cut both strands of DNA and could be dangerously inaccurate, to more precise and easier-to-use technologies to target genes in specific ways. Today, many companies are editing genes to realize benefits in farming, chemical manufacturing, medical treatments, and pest resistance.

Some users seized on it for making faster racehorses, hardier crops and feed animals, better pets, and tastier cheeses and liquors. CRISPR's effect was felt first in agriculture, making more productive cows, tastier tomatoes, and more nutritious rice. The natural next step would be to try and repair human genes, such as the single gene error that caused Victoria Gray's suffering. Around the world today, some forty clinical trials are using CRISPR to address diseases such as sickle cell anemia, heart disease, diabetes, cancer, hyper high cholesterol, hereditary blindness and other genetic diseases in institutes from Milan to Milwaukee, from Leiden in the Netherlands to Hangzhou Cancer Hospital in China. CRISPR was the closest thing to a revolutionary lab tool synthetic biology had yet offered.

From the beginning, however, the breakthrough offered a temptation. As CRISPR became available in kits marketed to the public, some people might use it to try to improve their appearance, muscle tone, and so on, sometimes recording the unfortunate results for upload to YouTube. Its enticement was too great to resist. One lab discovery cracked open a new world.

DIAMONDS IN THE ROUGH

Several attempts were made in the 1990s to edit human genes for medical therapies. In 1990, two children with severe combined immunodeficiency (SCID), or "bubble boy disease," were treated with a virus-based transporter of genetic material to bolster their immune systems. A trial in the United States succeeded.[5] In 1999, a seventeen-year-old boy, Jesse Gelsinger, died in a gene therapy trial to treat his rare metabolic disorder. The FDA investigation found lapses in adhering to the protocol and in the informed consent Gelsinger had signed. In a 2002 French trial, one of ten children receiving gene therapy developed leukemia. That trial was halted.[6] After such experiences, Berkeley's Jennifer Doudna noted: "The term gene therapy became a kind of black label."[7]

Growing up in Hawaii, Jennifer Doudna loved walking on Hilo Beach, watching the canoes race and the crabs scramble in the tidal pools. As a smart, tall, blond girl, Doudna felt like an outsider in school. "I loved math," she said. "I was a science nerd."[8] Noticing her interest, her English-professor father gave her a copy of James Watson's *The Double Helix* when she was eleven. The story of a high-stakes race to unlock DNA's beautiful structure entranced the young girl.

It was the double helix that gave the world biology's "central dogma": DNA is the template for RNA, and RNA is the template for making protein. DNA is the library of cell information, and RNA is the messenger that puts the information into action. In graduate school at Harvard, Doudna researched the structure of RNA in Nobel-winner Jack Szostak's lab, specializing in the cell structures called ribozymes, made of RNA but acting like proteins. Later, Doudna worked as a postdoc with Nobel-winner Tom Cech at the University of Colorado to better understand those ribozymes, which many scientists suggested offered clues to life's origin. Both mentors conducted significant research with practical applications, little knowing their talented graduate student would surpass them.

At the time, messenger RNA was not as well known as it would later become with SARS-CoV-2 (the RNA virus causing COVID-19) and the vaccines that combated it. Unlike DNA,

RNA typically has only one strand. It contains three of the same bases as DNA, A, C, and G, but replaces the fourth, T, with U (uracil). Its extra oxygen atom makes RNA an oxyribonucleic acid, unlike DNA which is a *de*oxyribonucleic acid, cumbersome name that it is. RNA comes in several guises including messenger RNA, a communicator, and transfer RNA, a cell guide. It is ephemeral, malleable, and universal, often used by viruses to hijack a body's defenses. RNA works quickly, then is discarded like a Snapchat message.

While she was working in Boulder, Doudna's father became sick with inoperable melanoma. She raced home to Hawaii to share his last few months, listening to Mozart, reading aloud Thoreau's *Walden*, and helping nurse him. After he died, a grieving Doudna accepted an assistant professorship at Berkeley, bringing with her a new husband, a baby, and a conviction that life was short so one must pursue only important science. While researching RNA interference, a technique that could neutralize disease-causing gene mutations, she got a call from a Berkeley biologist she admired but had never met.

The Australian-born Jill Banfield was studying ancient microbes in extreme habitats such as abandoned gold and silver mines of California and South Africa. These strange specks of life harbored a genetic mystery, Banfield told Doudna over tea one blustery spring day. Their microbial DNA had lengthy, repeated genetic palindromes, sequences that spelled almost the same way forward and backward, like "senile felines."[9] The same, seemingly useless sequences had been discovered in microorganisms in extreme habitats thousands of miles away from each other. It was as if the same nonsense words were inserted between sentences in obscure, different books from around the world.

Banfield drew on a napkin a series of diamonds, the palindromes, alternating with squares, the microbes' natural genes.

The palindromes proved to be gene sequences from viruses that attacked the bacteria. It was supposed that these palindromes showed evidence of a defense system whereby bacteria destroyed their invaders by cutting their DNA and storing the gene sequences end-to-end to be remembered in case of future attacks. But no one knew how that was done. That was what Banfield was asking Doudna to study.

Captivated, Doudna and others figured out that the cutting was done by CRISPR-associated (Cas) enzymes acting as scissors. There were two CRISPR systems, one requiring several enzymes, which Doudna worked on, and another simpler system, requiring only one enzyme called Cas9, which French scientist Emmanuelle Charpentier was studying. Like Banfield, Charpentier also wanted very much to meet Doudna.

A FASCINATING POTENTIAL

Born in 1968, Emmanuelle Charpentier was a lithe, determined researcher inspired by the great French chemist and microbiologist Louis Pasteur. Charpentier was investigating the bacterium *Streptococcus pyogenes*, which caused scarlet fever and could lead to such fatal infection it was nicknamed the "flesh eater." She uncovered the bacterium's CRISPR defense. Traveling to a job interview in the frigid north of Sweden, Charpentier heard the "crisp, crisp" sound of her footsteps in the frozen snow.[10] "It was a message, I felt, to continue my research," she said.[11] Her lab discovered that the enzyme Cas9 and a molecule in the cell, called transfer RNA, were critical elements that made the bacterium's CRISPR sequence work.[12] She was onto something big but needed Doudna's RNA expertise to unlock the mechanism. She planned to meet her at a conference in Puerto Rico.

They met by accident at a café and, walking through Old San Juan, they agreed to join forces. The CRISPR immune system protected bacteria against invading viruses called phages. The bacterium protected itself by cutting the phages' DNA and storing the pieces. It was a phenomenal discovery. But Doudna and Charpentier decided to take it further. What if scientists could learn, like the bacteria, how to cut and splice DNA themselves? That is what the two attempted to understand.

In Doudna's lab, at first nothing worked. Then they fused their CRISPR RNA with Charpentier's tracrRNA to make a simplified guide. To test it, they picked a known gene in the DNA in Doudna's freezer and picked five places where the gene could be cut. They changed their CRISPR guide-RNA sequence to match the targeted five sites and, to their delight, it worked. The enzyme cut the DNA every time in exactly the right place.

Doudna and Charpentier's paper describing the breakthrough was published in *Science* in an unprecedentedly fast twenty-eight days after they submitted it in 2012.[13] The news ignited the molecular biology world. Not only was it a superb insight in itself, but it also offered a powerful lab tool that could cut and paste genes like sentences in a story. Until that moment, a long-promised gene revolution had been slowed because previous techniques were so cumbersome. CRISPR was a platform. Dreams of editing plants and animals to make them more useful or of attacking genetic diseases like Huntington's or Tay-Sachs seemed within reach. The world's best geneticists rushed in.

FROM A TRICKLE TO A FLOOD

Initially, papers about experiments using CRISPR appeared at a normal pace, but quickly the pace accelerated. Researchers edited genes in cells as different as stem and leukemia and kidney cells, in organisms as varied as mice and zebrafish and microbes and more. CRISPR was so easy to use and reliable, a researcher could almost watch a YouTube video and then apply it. The tool was a step toward "kitification" that made synthetic biology techniques accessible to both experts and amateurs alike.

Within months, the Doudna-Charpentier paper was being cited at the astounding rate of once every eight hours. Companies with names such as GeneCopoeia, Intellia Therapeutics, Precision Biosciences, Editas Medicine, Horizon Discovery Biosciences, Sangamo Therapeutics, and more sprang up to sell editing kits or services. Doudna started several companies and Charpentier three of her own. Competitors at Harvard and the Broad Institute, who adapted CRISPR to human cells, were on their heels.

The possibilities seemed endless. One could edit microbes to make ingredients for vegetarian meats, chemicals for diabetes or leukemia treatments, ingredients for microbially manufac-tured fertilizers and fragrances, and even for building and cloth-ing materials. Horse and prize dog breeders ordered the kits for experiments in backyard shacks. Researchers applied this boon to address single gene diseases like sickle cell anemia, along with goals of attacking crop predators without pesticides, and improving nitrogen fixation by plants. By August 2013, a mouse with edited genes was born.[14]

But the overwhelming temptation was to treat human afflic-tion. Harvard's George Church and the Broad Institute and MIT's Feng Zhang showed that human cells could be edited by CRISPR, and they tried to work together with Charpentier and Doudna in developing the discovery for commercial and clini-cal purposes. At first, there was no rush. Investors were not ter-ribly interested in the years 2013 and 2014. Businesspeople kept confusing the technology's name with crisper foods.[15] One of the few, early, interested companies, though, was a little-known start-up called Moderna.

Already, people were concerned about the ethics of editing human genes, whether to prevent disease or to augment desirable traits. Researchers gathered in closed meetings, one in New York in 2015 and some 130 scientists in another in Cambridge, Massachusetts, in May 2016, to discuss guidelines for such research. The meetings were criticized for being closed to the public, and subsequent meetings were then opened.[16]

The patent fight was the next sign of trouble. As their collaboration cooled, the researchers, Church and Zhang, Doudna and Charpentier, raced to apply for patents on human gene editing techniques. The U.S. Patent and Trademark Office awarded the patent to the Broad Institute, which had paid for a rush order, making its bid the first to be reviewed. The Berkeley group appealed the ruling. The continuing patent issues had huge economic stakes. But Cas9 was not the only usable enzyme scissor. There were also Cas12, Cas14, CasX, and CasY. Companies like DowDuPont, Cellectis, and MilliporeSigma each won patents on the new scissors.

With so much at stake, the situation was a mess. It was about to get worse.

"COWBOY SCIENCE"

For centuries, humans have sought to enhance themselves, using everything from elixirs to electricity. When molecular biology opened the human gene door with recombinant DNA in the 1970s, a respected researcher, Maxine Singer, sounded an alarm. At the Asilomar Conference in California in 1975, some 150 biology leaders and four nonscientific participants had formulated the initial ethical guidelines for what was then called recombinant DNA, or early gene editing. The resulting guidelines, which we return to later in this book, strongly discouraged researchers from editing human germlines, the embryo genes passed to future generations.[17] The recommendations created the framework for grants rules and guidelines later adopted around the world.

But the wide availability of CRISPR kits brought a whole new level of challenges. CRISPR's power and precision kept improving, and ambitious researchers wanted to capitalize. While researchers edited genes in animals and plants, explorations of human gene editing quietly sped ahead.

One little-known researcher in the field was the Chinese national He Jiankui (JK), trained at Rice University, who taught at Shenzhen's Southern University of Science. In 2017, he began consulting both a Stanford gene sequencing expert, Stephen Quake, and a Stanford bioethicist, William Hurlbut, about acceptable ways to use gene editing to help humans. In 2017, he attended a small worldwide gene editing ethics conference assembled by Doudna at Berkeley, open to those qualified to attend. In the auditorium, heated arguments broke out about rules for applying the technique in medicine. JK told Arizona State University ethicist Ben Hurlbut, William's son, that he took to heart the comment of one researcher, "many major breakthroughs are driven by one or a couple of scientists. Cowboy science" is how this is going to get done.[18]

In late 2018 in Shenzhen, JK implanted gene-edited twin girl embryos, fertilized by an HIV-positive father, into their mother. He edited the *CCR5* gene, with the purpose of preventing the ability of the AIDS virus to infiltrate the girls' white blood cells. *CCR5* produces a protein that can serve as a receptor for the AIDS virus in the nucleus. If the gene was disabled, Harvard's George Church had once suggested, it might possibly prevent

AIDS infection in HIV-positive individuals.[19] In late October, gene-edited twin girls with the pseudonyms Lulu and Nana were born prematurely. The alteration of their DNA was a landmark breakthrough that JK announced to a packed auditorium at a November Hong Kong gene editing conference, accompanied by a slick video that was soon removed from the internet.

The science world erupted. There was no peer-reviewed paper, only the video and press release that said little about the procedure's details, or even the names of the doctors. Little data was offered to show the editing had actually achieved what it claimed. In previous human embryonic gene editing, off-target genes had been inadvertently altered. The fact that the father was HIV-positive, a condition with a strong social stigma in China, might have brought unfair pressure on the parents to consent. The edited genes conferred only a possible resistance to potential future infection later in life. The gene edits were not necessarily the right ones to confer such resistance to infection, which could be avoided anyway with safe sex practices. The announcement claimed the twins were born in November, but it was revealed they were born in October. There was evidence the gene edits were not uniform, and it was confirmed later that only one of the girls had a fully successful result, the other having only a partial gene modification. The funding source for the experiment was unclear. A third child, with a gene edit by JK, was also born in 2019.

In China, He Jiankui at first offered a defiant response. "They may not be the director of an ethics center quoted in the *New York Times*," he said of the parents, noting polls that showed 95 percent approval from HIV-positive Chinese patients for gene editing, "but are no less authorities on what's right and wrong—because it's their life."

As summer slipped into the fall of 2019, more negative press coverage and editorials appeared in journals ranging from *Science* and *Nature* to the British medical journal *Lancet* and the *Journal of the American Medical Association*. Global science organizations, including the World Health Organization, began developing a new set of guidelines for human gene editing. Several journals rejected JK's paper, which failed to identify the babies' primary care doctors, suggesting the doctors may not have been fully informed of the scope of his intervention. Condemnation came from leading media outlets like the *Washington Post, Stern*, and Fox News. JK was removed from his university position, and rumors claimed that he was placed under house arrest. The documents were sealed, and investigators learned little more.[20]

However, in April 2022, JK was released from prison.[21]

Elsewhere, however, gene hacking kits for pets and farm animals were available on the internet. Such editing was "really pretty easy," said one breeder.[22] From individuals around the world, with a variety of motives, arose a will to apply the new technique to helping people.

"I CONSIDER MYSELF CURED"

In her childhood, Victoria Gray loved computers and fishing with her father, playing basketball, and studying biology.[23] Some days she felt okay and pursued her dream of becoming a nurse. After all, she knew her way around hospitals from all her blood transfusions. She was doing well in college, with a dream of transferring to get her nursing degree. But she did not get better. She could not study and dropped out, giving up hope to become a nurse. Her sickle-shaped blood cells were damaging her heart. Many patients did not live past fifty. It was horrible knowing she could have a stroke at any time.

Sometimes she would be rushed out of the house in the middle of the night. Her oldest child, Jemarius, understood what was going on. He was sullen in school, getting into fights. His teacher told Gray, "Jemarius is acting out because he really believes you're going to die." Gray prayed and was looking into getting a bone marrow transplant, which was how she first came to Dr. Haydar Frangoul, medical director at HCA HealthCare's Sarah Cannon Research Institute Center in Nashville. Gray wanted to be evaluated for a transplant.

Frangoul mentioned a medical trial to edit the sickle cell gene with CRISPR. "We don't even know if it works. It's never been tried in humans," Frangoul said. "Do you want to try getting into the study?" he asked.

"Sign me up," Gray said.[24]

That spring, doctors at Nashville's HCA Healthcare began by removing Gray's bone marrow cells and editing them with CRISPR to get them to make fetal hemoglobin. In the searing summer of July 2019, lymphologist Haydar Frangoul reinserted billions of these cells into Gray's body. "The idea is that the fetal hemoglobin will take over the bad adult sickle hemoglobin cells," he told National Public Radio, which made a documentary about the procedure. Her father, Timothy Wright, came from Mississippi to help and keep her company. After several more months, she moved into an apartment.

Around the globe, CRISPR became a widely used, inexpensive lab tool that translated into tangible products created by fermentation of edited bacteria and yeast. These products included sugar substitutes from Cargill and other companies, milk whey produced by a company called Perfect Day, and other plant pathways producing steroids and painkillers. The fatty oil called squalene was re-created by gene-edited microbes to encase vaccines. Gene-edited mosquitoes were released in Brazil to prevent the spread of diseases. Synthetic biology was making organisms to change the world.

To be sure, some scientists distinguish gene editing from the bigger goals of synthetic biology. Gene editing alters nature while synthetic biology builds it anew. CRISPR is a tool, whereas synthetic biology is a broad, interdisciplinary field that includes topics we have covered so far (metabolic engineering and standardized parts) and those to come (directed evolution in chapter 4 and semisynthetic organisms in chapter 5). For our purposes, synthetic biology includes CRISPR gene editing as one of its crucial tools.[25]

In July 2021, the World Health Organization issued two reports guiding the use of CRISPR in humans. The reports came after three years of worldwide consultations with religious leaders, indigenous peoples, and patients' advocacy groups.[26] The first report was a working set of guidelines with case studies that echoed and expanded on the Asilomar Conference and the story of Nana and Lulu. It forbade embryonic gene editing. The second report offered nine categories of review and procedures for reporting abuses and called for an international registry of all gene therapy clinical trials, many of which are currently available in the United States on the website www.clinicaltrials.gov. How to enforce the rules, however, was not addressed. For the moment, the worst that could happen would be for a researcher to lose federal funding.

Gene therapy researchers were studying treatments for deafness, diabetes, pancreatic and other cancers, hemophilia, angina and non-small-cell lung cancer. Inherited eye disease was another target, involving edited viruses being injected into the retina, with few deleterious side effects.[27] Some researchers injected CRISPR-edited genes via viruses to treat chronic bladder infections and other diseases.

By 2022, technologies had improved. Safer and more accurate forms of gene editing that did not cut both DNA strands, but rather altered one or two bases or proteins, called base and primer editing, were some of the more sophisticated and precise tools coming into the laboratories.

Gene editing was saving lives and was being tested for ailments including X-linked adrenoleukodystrophy at Shenzhen Geno-immune Medical Institute. In Pasadena, California, it was helping Katherine Wilemon and her daughter avoid heart attacks caused by hypercholesterolemia.[28] One of the most-used techniques of synthetic biology, CRISPR helped to democratize a field, extending opportunities to smaller labs at a time when approved therapies, like Gray's, remained expensive.

An Argentinian company was using CRISPR to make faster, stronger racehorses. Kheiron Biotech edited a horse gene for the muscle hormone myostatin. A company in the Netherlands was making gluten-free wheat. A Canadian company had come up with more nutritious, farm-raised salmon, and one in Norway was making sterile salmon that grew faster than normal, with fewer diseases, but which could not procreate and spread their genes with unforeseen consequences in the wild. Companies in Brazil and Ireland used CRISPR to grow spicy tomatoes more cheaply than the ingredient spice capsaicin could be harvested naturally. A company in Norwich, Tropic Biosciences, was making CRISPR-edited sustainable, pest-resistant coffee.

Gene editing was also helping to develop biofuels including jet fuel. The company Synthetic Genomics was producing CRISPR-edited algae biofuels so successfully that it signed a contract with ExxonMobil to make 10,000 gallons by 2025.

For her part, leading up to spring 2022, Victoria Gray had not been hospitalized in the two-and-a-half years since her gene therapy. "Graduations, weddings, I never thought I would see any of those," Gray said to reporters. "I consider myself cured." Other sickle cell patients experienced similar results. But the patent fight continued. Charpentier and Doudna argued that Zhang and Church had not advanced materially on their discovery. The U.S. Patent Trial and Appeal Board awarded Zhang and Church "priority" in their granted patents for uses of the CRISPR system in animal cells, which covered humans. But the ruling also gave Charpentier and Doudna a right to one critical part of the CRISPR kit.[29] Most gene therapies remained too expensive for most patients. Better gene editing techniques were on the way, because CRISPR editing could cause unanticipated gene alterations in human embryos. As difficult as the science was, its application was more so.

PANDORA'S BOX

In the base of the towering gold statue of Athena on the Acropolis was a relief depicting the birth of Pandora. In Greek mythology, Pandora was the first woman on Earth, skilled in crafts and intelligent, created by the gods who each bequeathed to her one gift in a box she was told never to open. The brother of Prometheus fell in love with her and married her, but Zeus was still angry at her brother-in-law for stealing fire. One day, the curious Pandora fell to temptation and opened her box. Every malady of humans flew out—sickness, poverty, greed, envy, sadness, laziness, and anger. In panic, Pandora tried but failed to jam the box closed, but her failure was also rewarded. At the last second, out flew the last gift, that of hope.

The discovery of CRISPR gene editing offered hope for new ways to treat genetic diseases. At an African genetics center in Benin, CRISPR-Cas13 was used to make an inexpensive field coronavirus test. At Dermatology Research Associates in Los Angeles, a clinical trial was completed for the use of CRISPR gene editing to treat adults with acne. Boston-based biotech Vertex Pharmaceuticals, with CRISPR Therapeutics in Cambridge, Massachusetts, were treating diseases like sickle cell anemia and beta thalassemia.[30] The two companies, in addition to Bluebird Bio, were in clinical trials for a gene therapy for sickle cell anemia. Other companies exploring CRISPR medical applications included SinoGene Therapies, Beam Therapeutics, and pharmaceutical giant Novartis.

Beyond medicine, CRISPR was turned to making fatter pigs, hardier cows and rice that would release less warming methane, as well as crops and livestock that could withstand the effects of climate change. One key to avoiding the public outcry over genetically modified crops like Golden Rice, argued supporters, was that CRISPR editing did not transfer genes from one species to another, but rather enhanced the native genes already present. At the University of California, Berkeley, researchers worked to improve crop photosynthesis and increase the nighttime microbial carbon sequestration that gives topsoil its dark, rich fertility. Berkeley's Jill Banfield, along with a team including Doudna and Charpentier, edited rice soil microbiomes to make the crop emit less methane and resist the effects of flooding.[31] In Chile, the United Kingdom, Italy and Cuba, researchers produced a CRISPR-edited tomato high in vitamin D.[32]

Arising from basic research by Banfield, adapted by Jennifer Doudna and Emmanuelle Charpentier, the technique offered a gift of hope to people like Victoria Gray. CRISPR researchers set to work to cure inherited blindness and Huntington's disease, and others continued to work on sickle cell anemia. CRISPR played a key role in the success of several synthetic biology companies producing useful sustainable products in agriculture, medicine, and remediation. It would play a pivotal role in COVID-19 vaccine development.

For Victoria Gray, life was improving. Some four years after her operation, she felt healthier than at any time in memory. Her new blood cells seem to have overpowered the damaged ones. "We are thrilled," her doctor, Haydar Frangoul, told National Public Radio. The treatment came just in time. Her husband was deployed to Washington.[33] The country was suffering through its COVID-19 lockdown. Her great aunt and the pastor of her childhood church had died of the disease. Protests over the killing of George Floyd hit their hometown of Forest, Mississippi. Home alone, Gray would not have been able to cope with rearing her three children, who all noted the overall improvement in their mother's health The treatment was expensive, however.

CRISPR became a platform for programming life. It is a beautiful discovery and a lifesaving technique if used properly. It is also likely that, as techniques and oversight improve, more attempts to edit the genes of human beings are to come.

But then, another young woman glimpsed a way of manipulating a natural process to make products humans need, directing evolution itself.

4

THE SILK ROAD

Directing Evolution

I wanted to rewrite the code of life.

—Frances Arnold, Nobel Prize winner

At the University of California, San Francisco (UCSF), in 2006, a thirty-year-old chemical engineer named Chris Voigt was facing his critical third-year review. The trim, laconic researcher with a crew cut outlined a vision to "push the scale of genetic engineering" by programming cells to detect cancer tumors and other signs of disease.[1] But his bold ideas on modifying gut bacteria to attack cancer tumors seemed to draw little response. "It was the closest I've come to being fired," he later told an Office of Naval Research conference, smiling.[2]

He had been inspired by his PhD supervisor, Caltech's Frances Arnold, a former aerospace engineer captivated by the beauty and precision of nature. But engineering nature, say something as simple as a protein, was so incredibly difficult, Arnold once said, "It was terrifying."[3]

Frances Arnold had an answer. You do not need to engineer life from scratch. She had figured out a way to speed up the process of creating useful enzymes, especially the enzymes used in the chemical industry, detergents, or medicine. She mutated the genes that encoded the enzymes, then watched for new and improved versions to develop. Then she took the mutations for the improved versions and started the process again. She called it directed evolution.

Voigt wanted to take that idea further. He and others envisioned using computer algorithms to program complete cells and organisms, and more. The processes could be scaled up to industrial speed. In fact, he argued, cells could become living computers. The only trouble was, many critics did not believe in the idea.

But his students did. Companies founded by students trained in his and other labs were determined to take the next steps in synthetic biology. The companies would include names like Bolt Threads, making clothing; Asimov, engineering cells for therapies; Pivot Bio, creating a bacterium to replace energy-consuming fertilizer; and others.[4] On the other side of the world, the company LanzaTech in New Zealand was hoping to manufacture jet fuel from steel mill exhaust and modified ancient bacteria. Together, these researchers sought to design new, more useful organisms in a fraction of the time nature took, industrializing microbes to help make pest-resistant crops, clothing, fuels, and more.

To be sure, the technologies did not exist yet. To create those, scientists like Voigt's lab graduates might have thought back to the inspiration of Voigt's thesis advisor, Frances Arnold.

"THE LUNATIC FRINGE"

When Frances Arnold was growing up in Pittsburgh, she battled daily with four brothers. Her father was a nuclear physicist, her grandfather a World War II general. Born in 1956, she was an athletic, attractive woman who rebelled in high school and hitch-hiked to Washington to protest the Vietnam War. She rebelled so much that her parents gave her an ultimatum: straighten up or leave. She chose the latter and left home at age fourteen, working odd jobs, including waitressing at a jazz club called Walt Harper's Attic and driving a taxi, before entering Princeton and becoming one of the first women to major in engineering. After a gap year in Italy and Spain, where she drove a motorcycle from Milan to Istanbul, she went to the University of California, Berkeley, to pursue a doctorate in engineering.[5]

There, she became entranced by biology. "I was blown away by the beauty of living things," she recalled. She pursued her goal of directing the evolution of enzymes. Arnold came to evolution with an engineer's approach: take what works and improve it. She succeeded in evolving an enzyme that made a carbon silicon bond better than nature could create on its own, and worked on more catalysts to drive reactions that otherwise might take years to trigger. "Biology is the product of evolution," she would say, "so why not use evolution to engineer biology?"

Many colleagues doubted the approach. "It was considered the lunatic fringe," Arnold recalled. "Biologists did not do that. Gentlemen didn't do that. But since I'm an engineer and not a gentleman, I had no problem with that."[6]

Her designed enzymes could yield, she hoped, precursors for renewable fuels, biodegradable plastics, or even new drugs. The process made a notable pioneering achievement, one reviewer wrote, as if quoting Melville, "sent on through the wilderness of untried things to break a new path." But she was working mostly alone. "Frances essentially invented the field," Diana Kormos Buchwald, director of Caltech's Einstein Papers Project, later told the *New York Times*.[7]

Arnold studied enzymes, and then survived breast cancer, the chemotherapy for which "knocked out" her short-term memory," she recalled.[8] Despite the recognition of peers who elected her to the prestigious National Academy of Sciences, few outsiders knew of her work. That would change.

"SILK HAS A WAY OF SUCKING YOU IN"

The years 2004 to 2010 were a period of rapid growth in synthetic biology. "Whole genomes were being synthesized. DNA sequencing was coming down in price and going up in efficacy. The entire microbiome was starting to be pulled apart,"[9] Voigt told me. New techniques, such as RNA sequencing (RNA-Seq), allowed researchers for the first time to see the impact of their changes on the molecular composition of the cell.

At the Voigt lab in San Francisco in the years 2006 to 2009, two graduate students were trying to make biofuels. Gas prices were high and electric cars inefficient. On the fourth floor of the nondescript, new Beyers Hall in San Francisco's Mission Bay District, they struggled to get the bacterium *Salmonella* to make biofuels. If the shy, goateed computer engineer Ethan Mirsky and the voluble, green-eyed biologist Dan Widmaier could break down cellulose to be their feedstock, they would save a lot of money. But they had problems. The bacteria produced proteins that clumped together, clogging up the system.

They intentionally tried to produce several proteins, one of which was a spider silk protein. As gas prices fell and electric cars improved, the motivation for making biofuels slackened. The fact they made silk was "expected," Dan Widmaier recalled. "The clumping and problems inside the cell was what surprised us."

They considered trying to get the cells to secrete the protein. Voigt half-joked, "You should try to secrete silk, because if it's successful, you'll make headlines!"[10]

Widmaier had the silk secretion working, but not "nailed down enough" to get a paper accepted. That was when Northwestern University researcher Danielle Tullman-Ercek joined them. Twenty-eight-year-old Tullman-Ercek was used to striking out on her own, having grown up as an air force child. Back then, Merced, California, was little more than a U.S. Air Force base, but leaving that small town after first grade was so traumatic for her that she went from an outgoing child to a quiet one overnight, and remained shy until adulthood. Her family moved to Bossier City, Louisiana, then Maryland, and then too many other places to name. She had no idea of becoming a scientist until her senior year of high school. She learned to be self-reliant and independent, however, and found she loved chemical engineering, which she studied at the Illinois Institute of Technology as an undergraduate and then at the University of Texas at Austin for a PhD, where her mentor mentioned a new science called synthetic biology. Fascinated, she won a postdoctoral position in the Voigt lab to contribute her experience in the *Salmonella* secretion system, which lab members were lacking. She knew how to get proteins out of a cell.

The enticing idea of getting bacteria to make silk proteins had eluded many labs in the past. A Canadian firm tried and failed; a Japanese company did too. Adidas created a biodegradable silk shoe, then gave up because the market was so small. But spider silk was one of the strongest known materials: it threads thinner but twice as strong as that made by silkworms and was of keen interest to the military for many purposes. It could also be used for sutures, where silkworm silk caused an immune response.[11] Widmaier, Mirsky, and Tullman-Ercek decided to try.

Widmaier cloned DNA while Mirsky wrote computer code. Tullman-Ercek worked on secreting the silk protein from the cell. "It was this Cambrian explosion of ideas," Voigt later told *Forbes*. Then bioengineering PhD candidate David Breslauer, a tall, bespectacled contrarian raised in Oakland, contacted them. "I had been building devices and trying to dissect spiders, which gave a negligible amount of protein," he told me. "The engineering dean said there's a UCSF professor working on the recombinant production of spider silk. Call him!"[12]

Working together, they became caught up in their joint quest. "Silk has a way of sucking you in," Breslauer said. "We were coming with a lot of new technologies, DNA synthesis, microfluidics, all these things that were new on the scene."

Others were also trying to engineer spider silk, including a Washington University chemical engineer and a Michigan biotech company seeking to supply the U.S. Army with steel-strong parachutes. Even a microbrewery, Arachnid Ale, was using yeast to produce spider silk.[13] Others followed in later years by utilizing CRISPR gene editing or tapping MIT-based BioBricks' standardized biological parts list. The Berkeley silk makers were offbeat and persistent. "When other faculty are saying, 'he seems nuts,'" Voigt told a business reporter, "it's going to attract a student who is a little crazy."[14]

"TO OBTAIN THE IMPOSSIBLE, ONE MUST ATTEMPT THE ABSURD"

Across the world in New Zealand, a company was seeking to make jet fuel by modifying ancient microbes found at sea vents on the ocean bottom as well as in, of all places, rabbit droppings. Many had abandoned the biofuel dream, but plant geneticist

Sean Simpson was reading about microbes that could ferment carbon waste gases to make ethanol. "The fuels would be inherently sustainable," he said to me.[15] With long blond hair, Simpson looked a bit like Cervantes's character Don Quixote, the original impossible dreamer. Simpson came across a 1990s article showing that microbes called acetogens could ferment carbon oxide gases, like those found at the ocean bottom. His insight was to realize these strange organisms would also then ferment steel mill exhaust. The article, "Origins and Relationships of Industrial Solvent-Producing Clostridial Strains," suggested that the *Clostridium* strain could be engineered to produce industrial chemicals. What Simpson wanted was to use steel carbon gas to make ethanol. He named his company for the Spanish *lanza*, or lance, and won a government grant to build a plant in Auckland.[16]

Simpson focused on jet fuel because he thought electric cars would corner the automobile market, and "you cannot fly a jet on a battery," he told me. By 2009, LanzaTech finished its 15,000-gallon-a-year facility near New Zealand's only steel mill. It developed a proprietary microbe strain to be a biofuel workhorse. In 2010, it hired a seasoned renewables executive— Jennifer Holmgren—as CEO, who in 2014 transferred its main office to Skokie near Chicago and expanded teams in India and China. The company signed a joint venture with Chinese giant BaoSteel to process the waste from its Shanghai steel factory at demonstration scale.

Within a few years, the company expanded into Belgium, the United Kingdom, and elsewhere. Posing against the Pink Floyd album backdrop of London's famous Battersea power station, Holmgren and Virgin Atlantic's founder Richard Branson publicized their agreement to fuel a jet flight in 2018. It flew. Other airlines followed with flights running on sustainable aviation fuels (SAFs), including a December 1, 2021, United Airlines flight from Chicago to Washington, D.C., that flew fully on biofuel. A select few were making a go of synthetic biology as an industry.

"MAKE SURE YOU GET PAID"

For the silk makers in Berkeley, things were not going well, despite their success in harvesting genes from the orb-weaver spider *Araneus diadematus*. Their system had four parts—a genetic circuit; a unit of linked genes responsible for making their protein; a target; and a cleavage site to turn off the system. Silk protein was so difficult to engineer, "Smart people run away screaming," recalled Widmaier. The goal—spider silk itself—was strong, smooth, and soft. "Here's a protein that has this cool outcome where it goes from a liquid to an interesting fiber," Breslauer told me. "It hit all these buttons for me. It was this novel molecule that for some reason was beyond all our expectations and we didn't know why," he said. "But at first we were making salt mostly."[17]

Silk was so difficult to make that Widmaier and Mirsky became obsessed by their effort. Tullman-Ercek had the most experience with the pathway they were harnessing and with coaxing bacteria to secrete a protein through the cell membranes. Other systems would get a protein to a cell's wall, but not through and out. Finally they achieved yields good enough to publish. They built a DNA sequence from a computer database to get the silk protein gene. Rushing their paper to the journal *Molecular Systems Biology*, they announced, somewhat grandly: "This approach will revolutionize how natural diversity is explored when engineering cells."[18]

Some investors, like former scientist Rob Carlson, foresaw a new industry. With academic labs unequipped to produce synthetic biological products in quantity, many young participants from the SynBERC group started their own companies. The Voigt lab's silk seekers started their business in 2009. Two University of Iowa graduates, Karsten Temme and Alvin Tamsir, started Pivot Bio in 2010, a company to replace greenhouse-emitting fertilizer with programmed microbes to deliver nitrogen to crops. The Voigt lab graduate Elizabeth Clarke founded a company called Industrial Microbes in 2013, and another Voigt graduate, Alec Nielsen, created a cell programming company called Asimov in 2017. Tullman-Ercek preferred teaching. "I joke that I'm the only person from Chris Voigt's group not to have their own company," she laughed.[19]

Meanwhile, Mirsky and Widmaier took a course in entrepreneurship and, with Breslauer, decided to go for it. Calling their company Refactored Materials, they proposed a business plan to industrialize the microbial fermentation of proteins to make silk fibers that could be woven into material. They had little knowledge of how to run a business in 2009. "I thought you went and got an MBA," said Breslauer, whose father, aunt, and uncle were all scholars. "If this fails, I can always get a postdoc." Consulting his thesis advisor, Breslauer confessed, "Business development is not my jam."

"Whatever you do," his advisor replied, "make sure you get paid."[20]

"A LINE OF SILK"

At the time, a meeting was held at a historic San Francisco law firm named Orrick, where the key synthetic biology founders and funders gathered. The founders of Twist Bioscience, making synthetic DNA, were there but did not have a name or company. Zymergen's founders wanted to make high-speed automated tools to allow others to conduct research. They attended the meeting but also did not yet have a name. Ginkgo cofounder Jason Kelly attended. Ethan Mirsky and Dan Widmaier went. They had a company and a name, Refactored Materials, but were operating without any employees besides the founders.

The bacterium *Salmonella* was too inefficient to produce enough silk protein, they realized, and decided to start over. DuPont had shown that a yeast strain, *Pichia*, produced spider silk protein, and it had just come off patent. "We decided we would not take anything from graduate school, not any existing technology. We would build entirely new," Breslauer recalled to me. They applied for a U.S. Army grant to make super-strong silk for parachutes and clothing. The army was funding several synthetic biology start-ups to make materials and protect the health of its soldiers. Then they won another grant from the National Science Foundation.

At first, the entrepreneurs used an office at UCSF, where they ordered a group of *Nephila* spiders from a Florida dealer and let them spin webs from hula hoops suspended over the desks. They hoped to use the spiders to extract silk protein and test their spinning, Unfortunately, the spiders ate each other. One visiting molecular biologist popped in for a look and took off, yelling.

In their next location, a small incubator space leased from a Chinese company that had folded, Widmaier tried to clone the DNA for the desired protein. By 2014, he had purchased every cloning kit on the market, tapping the same technology used by crime investigators. None worked. He tried every combination. He tried every single condition. The indicator was a blank sheet he would wait to develop almost like a photo. Nothing showed up. The only thing that showed up was the control. They wondered if it would ever work.

One night, Breslauer walked into the incubator space and found Widmaier checking the experiment. "Oh, wow," Widmaier was saying.

"What?"

He showed Breslauer the paper. One line appeared, like a ghost. Breslauer's eyeglasses slipped from his face. "It was like, WHAT?" Breslauer said. "This line of silk polymer. Yeah, a line of silk."

They had a new name. Bolt Threads.

A ROUGH STRETCH

The following years were a rough period for many. In 2016, Frances Arnold lost one son in a car accident. The CRISPR patent lawsuits continued between the Broad Institute and the University of California over the lucrative use of gene editing in humans, worth anywhere from $100 million to $10 billion in licensing revenues.[21] Jennifer Doudna was approached by strangers with the idea to edit human genes. As the world's economy struggled after 2012, and with gas prices falling, some of the biofuels companies fired some workers. Other companies overpromised to attract investors. The technology was so new that the move from lab to factory production was challenging. The giant of synthetic biology companies, Amyris, teetered on bankruptcy.

At about the same time, the founders of Zymergen, Jed Dean, Joshua Hoffman, and Zachary Serber, the son of a

Columbia University professor, were running out of money. Serber loved Shakespeare and liked thinking about the role of science within the society that created it. He and Dean had met while working at Amyris and discussed that company's expensive strategy of having senior scientists or postdocs conduct experiments by hand. They wanted to automate synthetic biology on a huge scale and let discovery dictate what products could be sold. Their analysis suggested that some 15 percent of an oil barrel went to non-fuel product. They planned to create robotic labs to fabricate sustainable materials replacing that 15 percent.[22]

They had no customers. The bank account was zero. To economize, Serber lived with his wife and daughter on a houseboat in Sausalito Bay. "We did not pay ourselves for the first year," the tall, bearded, freckled Serber told me, earbuds hanging around his neck. "I probably should have done more research," he said. "Customers wanted evidence we had some capabilities." But without money, they could not show capabilities. "It was a Catch-22. We were meeting on a houseboat and we were at the end of our rope." Then they got their first customer, a Fortune 500 agricultural processing company.

A NEW SOLDIER

Back in 2014, a small group had assembled in a palm-enclosed conference center at Manhattan Beach, California, for the first annual Synthetic Biology: Engineering, Evolution & Design (SEED) meeting. It featured critical people like Columbia University Medical Center's Harris Wang, MIT's Velia Siciliano, and Danielle Tullman-Ercek, now with her own lab in Berkeley.

In the keynote address, MIT's Angela Belcher described directed evolution, "to select or evolve organisms to work with a more diverse set of building blocks." James Collins told the group, "Synthetic biology is bringing together engineers, physicists, and biologists to . . . construct biological circuits . . . and to use these circuits to rewire and reprogram organisms. These re-engineered organisms are going to change our lives."

In her office in the Air Force Research Laboratory near Dayton, Ohio, Biosciences Technical Advisor Dr. Nancy Kelley-Loughnane was following the field. The U.S. Navy was funding synthetic biology, as were the U.S. Army and DARPA. The Department of Defense had created a Biological Technologies Office and had funded early Ginkgo Bioworks research into infection prevention. The biggest funder of synthetic biology, aside from venture capital, was the military. The U.S. Air Force, however, was not yet a major player. Kelley-Loughnane wanted to change that.

Some researchers objected to taking Department of Defense money. They recalled the bioweapons programs of the 1960s, defunded by President Nixon in 1969. Some synthetic biology discoveries had dual uses. A medicine could be retooled as a killer. But the military was one of the strong early funders of synthetic biology research, for sensors, waste remediators, performance enhancers, and other non-weapons applications.

Then Kelly-Loughnane's phone rang. Her supervisor had just returned from the 2016 SEED conference in Chicago and was thinking engineered microbes could help the air force make materials, recycle waste, and transform its supply and waste service.[23] "We need to get a proposal out," her boss was saying.[24] It was what she had been arguing all along. It marked a turning point, but to what? No one was quite sure.

SILKY STRANDS

A copy of *Charlotte's Web* sat in the lobby. A pretty fountain gurgled outside the office by San Francisco Bay. The letters of the silk protein sequence covered a wall. At Bolt Threads' new digs in Emeryville, California, close to Keasling's institute and companies like Amyris, they had made a small amount of silk, tested it, and published it. "But if you're going to the large scale you have to make hundreds of fibers in parallel. That led to problems because we got this big mess," Breslauer explained to me. "The filaments stuck together. That was driving us crazy." As they struggled, they kept their profile low. Few people even knew how to contact them.

One day in 2015, Breslauer pulled the yarn fibers out to try a different set of conditions. The filaments separated beautifully and floated around, like a first step to a spider's web. "I practically broke into tears," he recalled.

Learning about their company years earlier, Steve Vassallo of Foundation Capital had sent them a LinkedIn message about visiting. In 2011, he led their first round of seed funding. More money followed in 2015 with a $32.3 million second round led by tech billionaire Peter Thiel. They signed a deal to provide material for Patagonia, and the company's chief commercial officer flew around the country to announce this was the biggest textile innovation since the invention of Gore-Tex. In the meantime, they installed sewing machines in an abandoned warehouse.

They had created silk, but silk with all kinds of qualities, and it was still hard to make it consistently clothing-ready. Eventually, Bolt Threads would switch from silk to leather made from mycelium, or mushroom roots. Still, when he got depressed,

Breslauer held the fibers—hundreds of white yarn filaments—long strands to make silk.

On the other side of the country, Ginkgo Bioworks was continuing to automate the new science.

"LOVE OUR MONSTERS"

From Boston, Massachusetts the Ginkgo Bioworks foundry, Bioworks1, was running and Bioworks2 almost completed. Their creative director, Christina Agapakis, wrote articles for the public with titles like "Smelling in Multiple Dimensions" and "Love Our Monsters—Radical Collaboration in a Post-Disciplinary Age." In the latter, she explored an idea raised by the science historian Bruno Latour in his 1991 essay "Love Our Monsters."[25] Rather than reject technologies like gene editing in favor of pristine or perfect nature, we could instead embrace the transformed creatures as children to be nurtured. These hybrid creations of synthetic biology had rights. Agapakis wrote articles focused on ways such designed life forms could create a more sustainable world. In May 2016, Ginkgo Bioworks CEO Jason Kelly published a *New York Times* editorial written by Agapakis, "I Run a GMO Company—And I Support GMO Labeling," noting the benefit of cheaper insulin made by genetically modified microbes in the life of Kelly's father, who had diabetes.[26]

By then, Voigt had moved from UCSF to MIT, where his team turned to creating a computer program in the DNA of bacteria. His lab created Cellular Logic (Cello) to help DNA compute, and made it available for free through MIT. He mused that they were close to building cellular computers on a par with *Apollo 11*'s primitive computing system, with its 5,600 logic gates. He and a few lab graduates cofounded the company

Asimov, publishing papers showing how such cell design could be streamlined and automated to make new medicines.[27]

Money trickled in, but the bioengineered consumer products were expensive and in short supply. In 2017, Bolt Threads had sold its first commercial product, a $300 necktie. The company sold fifty. Then it teamed up with English designer Stella McCartney to create bags, blouses, and pants. In 2018, it marketed a blended synthetic men's silk and wool "Cap of Courage" for $200 and sold one hundred. Yet within a year, Bolt Threads was valued at $700 million, and Ginkgo was valued at an astonishing $1.4 billion.

In early December 2018, Frances Arnold was in a deep sleep. The phone rang at 4:00 a.m. in her Dallas hotel room, near where she was scheduled to give a talk that day. She assumed it was a prank call, but reached out to answer. "You've won the Nobel Prize," the caller said, and requested her not to call home until the news was announced. Heart pounding, she jumped out of bed, "surprised, terrified, bouncing off the walls," she recalled. When the announcement finally went public, she called home. No one picked up.

"Frances Arnold pioneered the directed evolution of enzymes," the Nobel Committee proclaimed. "We call them fantastic bio-catalysts." Arnold's father beamed with pride. Suddenly people were paying attention, but a huge challenge still remained. The route from discovery to industrial-scale production was long. To witness a company making that journey "across the valley of death," I traveled to Skokie, Illinois.

"CROSSING THE VALLEY OF DEATH"

On a brilliant, hot summer morning in 2019, I drove my diesel Volkswagen to visit LanzaTech just north of Chicago, hoping its process might one day mitigate the exhaust of cars like mine. A glass-walled conference room was packed with people as I put on my lab coat, carbon monoxide detector, and safety glasses. "Be inquisitive" a wall stencil urged me. Michael Köpke, vice president of synthetic biology, pointed out the compressed gas tanks outside the window, guiding me past giant steel fermenter tanks, screens displaying gene sequences in computer algorithms, gleaming gene sequencers, and gas pipes labeled as carbon monoxide, hydrogen, and other gases as in a science-fiction movie. Tall and lanky, with blond dreadlocks pulled into a ponytail like his favorite reggae singers, Hamburg-born Köpke greeted workers as we passed through Fermentation Rooms I and II, and then into the crowded molecular biology room.[28]

The company had a commercial plant in China making jet fuel from steel waste gases, with plans for another, and partnerships with Dow and BASF for carbon recycling into other chemicals and with European partners to convert plastic waste into super-materials for products such as auto bodies, packaging, and textiles. The fuel its plants replaced, it estimated, was equivalent to taking 70,000 cars off the road. By then, company leadership had flown on a jet from Orlando to London fueled in part by ethanol that was the product of microorganisms fed with industrial waste. The company was on the brink, but making the transition from research to industrial production was like "crossing the valley of death," Köpke told me.

By 2019, synthetic biology was bringing together engineers, physicists, and biologists to model, design, and construct biological circuits out of proteins, genes, and other bits of DNA and to use the circuits to rewire microbes for producing fuels. Similar centers were growing in China, India, England, and the Netherlands. In the LanzaTech lab, the reengineered ancient organisms swirled and swished like milky clouds in the fermenters, with robots performing thousands of automated searches called screens all at once. The company was expanding its tools, with new collaborations with the Joint BioEnergy Institute and Northwestern University. It was trying to turn pine waste into fuels in Soperton, Georgia. The first of its two China plants had already produced several million gallons of ethanol, and the capacity gradually increased.

Other companies were leaping on the directed evolution techniques that Frances Arnold pioneered. Arnold herself cofounded three companies using directed evolution to create enzymes churning out valuable chemicals. One, Gevo, made biofuels. Another, Provivi, created nontoxic pest control, and a third, Aralez Bio, made unnatural amino acids as ingredients for medicines.[29] The companies Codexis, Direvo, and DIVERSA were racing to tailor enzymes for industrial uses. In the next step in directed evolution, newer researchers at Boston University and UCLA created continuous evolution technologies "to effectively make a steering wheel or guidance system for hypermutation systems," Boston University's Mo Khalil told me.[30] These new technologies, including one developed by Khalil, enabled much faster, more targeted evolution of useful enzymes in multiple parallel fermentation tanks.

Sustainable aviation fuels were being made from wood mill waste and, even, beef fat called tallow. At Eagle Country Regional Airport in Gypsum, Colorado, some of the jets at the Vail Valley Jet Center in 2022 were burning fuel made from 30 percent beef tallow and 70 percent kerosene, marketed to reduce a jet flight's carbon footprint by 25 percent.[31]

At LanzaTech as I left from that first visit, I saw road and mountain bicycles leaning against a lab wall, under windows looking out over waste gas tanks, a wooded bike path, and a church spire. Synthetic biology still seemed to ride on a hope and a prayer. But it was moving forward. Somewhere amidst the bike path's swaying switchgrass and Queen Anne's lace flowers, I imagined an orb spider, patiently weaving its web.

The field was making a web of connected insights and people. One of the last places to look was at remaking life's information molecules, DNA and RNA themselves.

5

WILD

Remaking Life

The first self-replicating cell on the planet that's parent is a computer.

—J. Craig Venter

On a cool, sunny April morning in 2022, I arrive in my cab by the YMCA in South San Francisco. A giant Genentech employee bus rolls by on Gateway Boulevard. A sign proclaims I am at the office of Twist Bioscience, the 950-employee gene supplier to the world, where I am meeting the smart and funny Jacqueline Fidanza, vice president of operations, and tall, bearded James Diggans, head of bioinformatics and biosecurity, for a tour of one of the world's foremost DNA makers.[1]

Researchers around the world were engineering DNA in microbes to make chemicals for marketable products. In Iowa, Renewable Energy Group was generating power from municipal waste. In Montreal, the company Enerkem made chemicals from engineered *E. coli*. In Pune, India, Praj Industries generated water and beer from waste. In Alameda, California, Industrial Microbes was making microbes to remediate waste.

In Paris, Eligo Bioscience was editing bacteria to make a product that neutralized acne on teenagers' faces. At Twist Bioscience, researchers ramped up the speed of DNA creation a thousandfold to help such labs make ingredients for medicines, clothing, food, and fuels. The company was building a state-of-the-art facility in Portland, Oregon. I had come to South San Francisco see how DNA was made.

But what if you could do better than merely making DNA? What if you could change life's four-letter code altogether? Was there something preordained about the double helix structure or was life's molecule a "frozen accident," as Nobel-winner Francis Crick wrote?[2] Can life be different? In Gainesville, Florida, a researcher named Steven Benner was creating new DNA with greater coding power than nature had given it. So were others in Berlin and Singapore, La Jolla, California, Gaithersburg, Maryland, and elsewhere. "I tell schoolchildren, it's like Legos," Angela Bitting, Twist's senior vice president for corporate affairs, once told me. Must life have only four blocks?

With more building blocks, you might expand DNA's ability to make useful products. One keen interest was in DNA as a digital storage medium, with greater capacity and longer shelf-life than silicon. Building synthetic DNA was "a grand challenge at the interface between biological, mathematical, computer and physical sciences and engineering,"[3] the National Science Foundation noted. But the quest had been going on for a long time, and promising results seemed far away.

Then new technologies such as inexpensive gene synthesis at places like Twist, acoustic microfluidics to move genetic material at places like Labcyte, and speedier automated sequencing at the J. Craig Venter Institute combined to pick up the field. By 2016, geneticist-turned-synthetic-biologist J. Craig Venter created two bacteria with synthetic genomes that journalists dubbed Synthia 1.0 and 3.0. When asked why, he said "because they told me I couldn't."[4]

In San Diego, California, the company Synthorx created semisynthetic organisms to make cancer therapies. Several international research collaborations, including one in the United States and one in Europe, accelerated the race to create alternate versions of life.

Others looked instead at RNA, the "first molecule of life" in the words of immunologist Ugur Sahin of Mainz, Germany—a molecule that could empower the immune systems of humans.[5] Taken together, the synthetic genetic efforts made an extraordinary quest to transform DNA and RNA and, in so doing, medicine, industry, and agriculture. How could the abilities of genetic material be expanded and tapped? To answer that, one first had to look at the past.

A BRIEF HISTORY OF THE SYNTHETIC GENE

Ancient myths are filled with half-human–half-beast chimeras, or beings possessing abilities that humans lack. Aristotle placed synthetic life forms under the term *techne*, or "art," which gave us the word *technology* and which, he felt, could never be as beautiful as the nature it imitated. Virgil offered a recipe for making synthetic bees. By the Middle Ages, however, *techne* came under assault as the church considered it to be a form of magic. The twentieth century's production of chemicals in engineered bacteria and yeast showed the promise of modifying life by moving edited DNA from one organism to another. With genetic engineering, chimeras made their comeback.

In 1971, as a start, Stanford chemist Paul Berg was the first researcher to transplant a gene, by splicing a bit of a bacterial virus into another virus that infected monkeys. This key accomplishment showed that viral DNA fragments could be combined and transplanted. But Berg did not move his foreign DNA into living organisms, because of public controversy about the dangers of the research.

A dozen years later, another researcher tried to change the nature of DNA itself. As early as 1989, at the Swiss Federal Institute of Technology, a blond-haired, dyspeptic, imaginative researcher named Steven Benner modified two of the four components of DNA, cytosine and guanine, and managed to get his synthesized versions to encode messenger RNA and thus make proteins. He wondered if DNA was the best molecule for encoding genetic information. In the following years, the effort expanded to get synthetic DNA into living organisms. The question was how.

In 2006 at Japan's Riken Genomics Sciences Center, researcher Ichiro Hirao made a completely synthetic, man-made piece of DNA replicate itself, showing that synthetic genetic material could perform many of the functions found in nature. Hirao moved on to an institute in Singapore, where he continued trying to add genetic letters to DNA, with an eye to crafting medical markers for disease detection and more.

Transferring to a University of Florida professorship, Benner made tiny nanostructures from his synthetic DNA. In 2005 he had created the company, Firebird Biomolecular Sciences, to commercialize his discoveries, marketing synthetic genetic material for diagnostic kits for HIV and hepatitis C infections. During the COVID-19 pandemic, Firebird would ship 20 million disease tests a month.[6] Step by step, various labs successfully made synthetic DNA translate into RNA and then to protein.

Then researchers decided the best way to discover the first principles of life was to find out how simple a cell could be.

THE QUEEN'S GAMBIT

In the early 2000s at the gleaming quarters of the J. Craig Venter Institute (JCVI) in Rockville, Maryland, a group of researchers led by John Glass was working to replace the entire genome of a microbe with that of another. Struggling to make a whole bacterial genome of hundreds of thousands of base pairs, the researchers finally created a synthetic DNA molecule twenty times larger than any made to date. But they could not get it to "boot up" inside the pathogen. It seemed to be an issue not with their technology, but with their target organism.

Researcher Carole Lartigue, who had recently received her doctorate in France, joined the JCVI team in 2004 to help solve the problem of eliminating a bacterium's native genome and replacing it with a new one that had been chemically synthesized. In France, she had worked on two simple goat pathogens, *Mycoplasma mycoides* and *Mycoplasma capricolum*, and convinced the team to switch targets and transplant the genome of *M. mycoides* into *M. capricolum*. She persevered but kept hitting dead ends. After two years of frustration, she walked into Glass's office, saying, "I can't do this. I'm out of ideas. I don't know what to do. I want to go home," Glass recollected.[7] The next day, Lartigue found two living colonies of bacteria with transplanted genomes. She had succeeded in building the world's first synthetic life forms.

Spending some $40 million in the quest, they had taken the genome of one bacterial species and inserted it into another. In a July 2010 issue of *Science*, they announced their creation, dubbed Synthia 1.0, had created proteins made by its inserted DNA. Venter's team demonstrated that life's instructions could be swapped and reprogrammed like a software update.[8] "Life's essence is information," declared *The Economist*. Still, other than the insertion of several DNA watermarks, the organism's genome was almost identical to that of natural *M. mycoides*.[9]

A series of efforts followed to modify bacteria for "toxic waste cleanup and energy production," Venter wrote. In his book *Life at the Speed of Light*, Venter predicted an industrial revolution that was based on products made by such modified organisms. "The goal is to replace the entire petrochemical industry," he vowed. To get to the point of adding new DNA efficiently, the first step would be to make a stripped-down bacterium. For the next six years, his team struggled to clear Synthia 1.0 of nonessential genes to produce what they called a minimal cell, showing the world the essence of life, they argued.[10]

Others approached engineering DNA from a different direction. On the West Coast in 2014, a small group of scientists at the Scripps Research Institute, who were led by a chemist named Floyd Romesberg, added a man-made DNA base pair they called X-Y onto the normal A, C, T, and G of an *E. coli* bacterium, which used the synthetic DNA to make RNA and protein and to reproduce itself, marking the first time anyone had expanded the genetic alphabet. "I was a chemist," Romesberg told me of his approach. "I thought the shape of the molecule shaped what life could do."

At JCVI, the team set about building the minimal bacterial cell they had long aspired to create. The plan was to precisely determine which *M. mycoides* genes were essential for life and which could be removed. At a team meeting, Venter made an unexpected appearance. He asked if anyone had put together a list of essential genes based on their preliminary data.

"Yeah. We have a list," Glass replied.

"Okay, let me have it, because we're going to make a Hail Mary genome."

The impetus for this "high-risk gambit," Glass recalled to me, was that Venter's new DNA company had excess capacity he wanted to use. Only unlike in football where a Hail Mary pass is thrown at game's end, they did it at the beginning. The effort failed but taught them key lessons.

After years of struggle, they finally succeeded. Described in *Science* in March 2016, their organism, "Synthia 3.0," had only 473 genes, far fewer than the thousands of any natural bacterium.[11] The basic parts list of life was hundreds "but not thousands" of genes, observed biophysicist Marileen Dogterom of the Netherlands, meaning life may not be as complicated as feared. "That's very exciting," she said. One revelation was that 149 of the essential genes had completely unknown functions.[12] "We don't know a full one-third of the basic knowledge of life," Venter marveled to reporters.

These efforts were building a new understanding of how to manipulate life's instruction book. If it could be simplified, that suggested it could also be expanded. If DNA was expandable, one could design an "uncountable number of applications," said Floyd Romesberg. The only way to tap that potential was to build cells with more fully synthetic DNA.

"THE AGE OF SEMISYNTHETIC LIFE IS HERE"

Inspired in part by Venter's accomplishments, Floyd Romesberg in California and Steven Benner in Florida redoubled their efforts to meet a "bigger challenge," Benner recalled. "If you go to Mars and find life, what is the chance that it will have any DNA at all?" Along with Hirao in Singapore and Petra Schwille in Germany, they tried to build completely synthetic DNA. Was there something special about adenine, cytosine, guanine, and thymine? With a grant in hand, Brenner eventually left his full professorship and its teaching duties to create his own institute in Alachua, Florida, the Foundation for Applied Molecular Evolution (FfAME), to devote himself full-time to the search. "I don't think there's any limit," Benner told me. "If you re-run billions of years of evolution, you could come up with a completely different genetic system with exactly the same properties."

Scientists had been working on altered DNA for years. With grants from NASA and the National Science Foundation, as well as the commercial rights to his early breakthroughs, Steven Benner in 2019 tried to insert four extra man-made nucleotides, two more base pairs, into DNA. His team began having success with doubling the number of bases in a molecule that formed a double helix. They tweaked viral RNA, the messenger, to get it to read the eight-base DNA. Expanding DNA's structure, they argued, would enhance its ability to make chemicals for fuels and goods. They were trying to "control Darwinian evolution in the lab," Benner said.

In Florida in the Benner lab, their eight-base artificial DNA triggered RNA to turn on a protein. Under the microscope the clear liquid lit up green, which meant the protein, tagged with a fluorescent marker, had been produced. They called their creation *hachimoji*: *moji* in Japanese for "letter," as in the term emoji, and *hachi* for "eight." It was "spectacular, a landmark achievement," said Romesberg, who had left Scripps and was working at his own biotech company. NASA said the discovery would "result in a more inclusive and therefore more effective search for life beyond Earth."[13] They had increased DNA's coding power.

Michael Jewett, professor of chemical and biological engineering at Northwestern University, called it "a true engineering feat (that) elegantly increases the number of DNA and RNA building blocks and dramatically expands the information density of nucleic acids."[14]

Romesberg's company, Synthorx (for "synthetic organisms"), was creating protein-based immune suppressors for cancer therapies and was expanding rapidly. Synthorx had a seemingly successful cancer candidate called THOR-707, and in December 2019 the French pharmaceutical giant Sanofi bought the small biotech firm for a sizeable $2.5 billion. "The age of semisynthetic life is here," Romesberg declared.

One little-known researcher from Hungary was also thinking about how biology could prevent disease, not by changing DNA as other companies were trying, but by modifying RNA, the cell messenger. She outlined her ideas in grant proposals and conference talks, but few paid attention.

"YEAH, YEAH. I CAN DO IT"

One of history's most fateful conversations occurred at a university photocopy machine in 1998. Hungarian-born biochemist Katalin Karikó introduced herself to HIV researcher Drew Weissman of the University of Pennsylvania, and they chatted about their work. "I am an RNA scientist," she told him. "I can make anything with messenger RNA." When virologist Weissman confided to her his pursuit of a vaccine for HIV infection, Karikó boldly replied, "Yeah, yeah. I can do it."[15]

Karikó's confidence may have stemmed from decades of overcoming challenges. After earning her PhD in biochemistry from the University of Szeged in Hungary, thirty-year-old Karikó immigrated to Philadelphia with her husband, Bela, and their daughter, Susan. As if anticipating the hardships that lay ahead, Karikó resourcefully stuffed $1,200 into her daughter's teddy bear to evade currency restrictions.

In 1989, she became an adjunct professor at the University of Pennsylvania and one year later began research on messenger RNA (mRNA for short). Her vision was to modify mRNA for therapeutic purposes. She believed that changing the genetic code of the universal messenger of cells could instruct the body to make its own medicine. If successful, synthesized mRNA could be the key to new vaccines, cancer and other disease medicines, and therapeutic agents to repair heart tissue. Most of the scientific community ignored or laughed at her. RNA was unstable and difficult to handle. While fellow researchers conducting conventional experiments won grants, Karikó received countless rejections and an eventual demotion at Penn.

She confronted both her own cancer scare and Bela's forced return to Hungary, all while working late nights and weekends on her mRNA idea. On the basis of the amount of time she spent in the lab, Bela calculated her wages amounted to about $1 per hour. "You're not going to work, you're going to have fun," he teased her. As University of Wisconsin researchers showed they could program mRNA in mice, Karikó persisted with her experiments, finally achieving a promising outcome. The detection by a dot matrix printer of all things, attached to a gamma counter to detect radioactive molecules, confirmed that new proteins were being produced by cells that would not have occurred without inserting her modified mRNA. "I felt like a God," she recalled later to reporters.

Soon after this success came the photocopy meeting with Weissman, and the two University of Pennsylvania researchers set out to make an HIV vaccine.[16] The trouble was, every time they tried to introduce the new mRNA into human cells, the body rejected it with a huge immune response. It took multiple tries and years of filling her prized lab notebooks before she and Weissman figured out how to make the body accept the modified genetic material. They added a molecule called pseudouridine, which allowed the synthetic mRNA to sneak into cells without causing a negative reaction.

In 2005 they published their success, claiming it could "give future directions into the design of therapeutic RNAs," and founded a small company called RNARx in Rydal, Pennsylvania.[17] But failing to achieve an agreement with them, the University of Pennsylvania sold the valuable license to a Madison, Wisconsin, company, Cellscript, for $300,000.[18] Two companies then approached Cellscript for rights to the mRNA technique.[19] One was a Cambridge, Massachusetts, start-up, then called ModeRNA, and the other was based in Mainz, Germany, and named BioNTech.

GOLDEN-WINGED MESSENGER

Mainz is the capital of the Rhineland-Palatinate, known for a picturesque old center with narrow lanes dominated by the half-timber, half-red-brick cathedral, and as the site where Guttenberg printed his early bibles that changed the world. There an immigrant Turkish medical couple, successful professors and also biotech entrepreneurs, had followed in the 2010s much the same avenue as Karikó and Weissman: learning to use mRNA for medical therapies.

Özlem Türeci had been a smart young medical student who grew up in Germany as the daughter of a Turkish doctor. She loved medicine and, in the oncology ward, met a tall, supportive,

dark-haired professor, immunologist Ugur Sahin. They both
felt frustrated by the limited therapies for treating their cancer patients and driven to research new ideas to improve care.
Sahin's father came to Germany as part of its 1950s Gastarbeiter
or "guest worker" program. His father toiled in a Ford car factory. Türeci and Sahin fell in love.[20] On their wedding day, the
guests being from their research group, they returned to the lab
to work. "We don't see what we do as work," Sahin said. "It's
really a way of life."[21] They founded and sold a successful cancer
therapy company, Ganymed. Then they thought about messenger RNA.

Single-stranded mRNA acts as a winged messenger carrying
the genetic code from DNA to the cell's ribosomes. The ribosomes
"read" the code from the mRNA and follow the instructions to
create proteins. The cell then "expresses" the proteins to commence
their designed functions. Messenger RNA is important in determining the body's response to various conditions. Among other
things, it could teach the body's immune system how to combat
new infections without using a live or weakened real virus or
permanently changing DNA. Because RNA is involved in every
protein a cell makes, if one could modify it, one could create a single easy-to-use platform applicable to treating multiple diseases,
simply by swapping out the DNA instructions.

Türeci and Sahin had turned to mRNA previously as the
most effective method to achieve their goal of personalized cancer treatments.[22] In Mainz in 2008 they had founded BioNTech
to create such individualized cancer treatments, focusing on
modifying the genetic messenger. The idea was to sample a patent's cancer tumor and design an RNA-based medicine for that
specific tumor. The company was also working on engineering
cells with mRNA to develop antibodies to other diseases.

Things were going well; they had some twenty possible candidates in the pipeline. In 2013 they recruited Karikó and assigned her to work on mRNA vaccines. Results were so promising that in 2018, they partnered with biomedical giant Pfizer to pursue a flu vaccine with the new technology.[23]

All the while, efforts to create new kinds of DNA continued.

EDITING NATURE AS MEDICAL THERAPY

While the British Medical Research Council's Jason Chin sought to reprogram the genetic code of life, the J. Craig Venter Institute was also trying to increase the powers of DNA. The plan was to translate the work from the genetics lab into the clinic. By 2019, JCVI had moved on from its initial quixotic pursuit of a minimal genome, derided by some critics as a gimmick, to develop needle-free treatments for type 1 diabetes by modifying microbes to help the body make its own insulin. That same year, scientists at the Swiss public research university ETH Zurich created the first bacterium designed entirely on a computer, *C. ethensis-2.0*. In 2020, JCVI successfully created its insulin-secreting gut bacterium for diabetes, and progress accelerated.

In San Diego, Synthorx was forging ahead with its anti-cancer therapeutic. Phase II safety trials were successful. That meant proteins made with unnatural amino acids by an unnatural base pair "were actually in people with cancer," Romesberg told me. "This was real. We went from a neat story in the lab to something that may actually help people." That was why Synthorx attracted the attention of international pharmaceutical company Sanofi.[24]

Along the way, some of the semisynthetic organisms' other flaws were addressed. One problem with Venter's stripped-down organism was that its offspring were of bizarre shapes and sizes. This was solved in 2021 by Elizabeth Strychalski of the National Institute of Standards and Technology. She added seven genes to make the cells divide uniformly, bringing researchers closer to "engineering fully designed . . . and controllable organisms," observed the University of Minnesota's Kate Adamala.

The accelerating speed of discovery was due in part to lowering prices of synthesized DNA. Companies like IDT in Iowa (discussed in chapter 2), Blue Heron in Washington State, GenScript in New Jersey, and Twist Bioscience in San Francisco, California, were helping to industrialize genetic research. I decided I had to visit.

TWIST

Based in South San Francisco, Twist Bioscience, the high-speed, automated producer of DNA, was expanding rapidly in the period 2018–2021. It was directed by the driven gene pioneer Emily Leproust, a French-born researcher trained at both the University of Houston and the life sciences and diagnostic company Agilent. Founded in 2013 by Leproust, fluid mechanics researcher Bill Peck, and data expert Bill Banyai, the company was named for the signature shape of DNA. Twist's innovation was to use semiconductor technology and a silicon platform to print thousands of DNA snippets (called "oligos") at a time on a silicon chip. The company offered improved accuracy, as well as lower prices and faster speeds than many competitors. In a partnership with Microsoft and funding from the Defense Advanced Research Projects Agency, it also led the effort to turn DNA into a medium for data storage, hoping to replace silicon.

Leproust was a born entrepreneur who at age twelve convinced her father to pay her a commission on VCRs she sold at his store.[25] "Build a new search engine, who cares?" she once told the young members of a conference audience in her sweatshirt and colored sneakers. "This is about making the world a better place."[26] She worked at Agilent Scientific Instruments for thirteen years, where she headed a division and learned the technical steps of making DNA. Coming from a family of business owners (her uncle ran a worldwide import-export company), she brought to Twist an understanding of customer needs as well as skills in team leadership and entrepreneurial methods. When Banyai and Peck developed their technology to write DNA on silicon using microprocessor technology, they needed a brash and confident CEO. Both thought of her.

Twist's key improvement was taking DNA chemistry, miniaturizing it onto a chip, and accelerating its production. Until Twist, the gene synthesis leader was IDT in Coralville, Iowa, which had been sold to Danaher Corporation in 2018 for $1.9 billion. Another company, out of JCVI, sold both gene fragments and a DNA printer. More such companies followed, such as GenScript, Genewiz, and Eurofins, to the point where gene factories recombining DNA snippets were a bona fide industry by 2020.

Companies like IDT put a single gene on a plate the size of a cell phone. Twist could fit 10,000 genes on the same plate, Angela Bitting told me. That enabled Twist and its customers to expand their DNA libraries. "We use our libraries to identify new therapeutics for our partners and our internal programs, and also have programs for storing data in DNA,"[27] Bitting explained. With its rapid delivery, high quality, and low prices,

Twist quickly went public in 2018, and by 2022 its valuation exceeded $2 billion.[28]

With Diggans and Fidanza guiding, I began on the first-floor silicon and DNA writer labs, where small DNA rings were assembled under dark amber light to prevent disturbance of their delicate structure. On the second floor, we viewed the rooms where the rings were assembled into genes, another where Twist made its own Petri dishes and equipment, and incubator rooms where *E. coli* containing the new DNA reproduced, making more copies rapidly. The company was on a roll. "We've grown so fast we took James's desk for a lab bench," Fidanza said.

Such power led some synthetic biologists to fear engineered viruses escaping labs. It was the job of Florida-raised Diggans, who had just been elected chair of the International Gene Synthesis Consortium, to protect the company's biosecurity by screening customers seeking pathogenic DNA for weapons. "I communicate daily with government agencies like DOD, Department of Homeland Security, Health and Human Services, and the Department of Commerce where I serve on the committee on biomanufacturing export control. I triage biosecurity," he told me.

Another major idea was to use DNA to store digital data. DNA lasted thousands of years, where silicon decomposed in perhaps a hundred. To test the idea, Twist joined with Microsoft and the University of Washington to store in DNA the data of the Deep Purple song "Smoke on the Water," and the Miles Davis classic "Tutu." Each musical note was represented by a unique triplet of the four bases (A, T, G, C) that comprise the building blocks of DNA. "Smoke" translated into GACCGACGTCAGAGC.[29] The two Bills and Emily Leproust kept expanding the company. In the same years, another company was also working on Karikó's messenger.

"BIG PROMISES"

On Technology Square in Cambridge, Massachusetts, Mode-RNA Therapeutics began with grants from the Defense Advanced Research Projects Agency and the Biomedical Advanced Research and Development Authority. It was connected to Karikó in two ways. One was through its cofounder, MIT chemical engineer Robert Langer, investor in or cofounder of some thirty previous biotech companies. Langer liked her theory that messenger RNA could be tweaked to help the body confront viruses. The other connection was ModeRNA cofounder Derrick Rossi, a Toronto-born biologist with signature huge red glasses and a soul patch, who thought Karikó's 2005 paper would win her a Nobel Prize. Rossi cofounded ModeRNA Therapeutics in 2010, along with Langer, and others.[30]

In 2011, the hard-driving French pharmaceutical manager Stéphane Bancel was hired, and in 2013, the name of the company was changed to Moderna LLC. Moderna researched treatments for heart and immune diseases but, after a number of failures, it switched its focus to vaccines. The advantage of mRNA for vaccines was that once it was unlocked for one virus, it could speedily be unlocked for others, in contrast to the years and years of work normally required to make a vaccine for each new disease.[31]

Investors placed their bets: $100 million in funding for research poured in, followed by an immuno-oncology partnership with AstraZeneca. "There were a lot of really big promises made," former Moderna chemist Jason Schrum once commented.[32] Gradually, some progress happened. In 2016, Moderna had its first human candidate for its mRNA-1851 vaccine, designed to protect against a flu strain. It conducted human trials for a Zika virus vaccine, a feared killer because it causes brain damage in infants born to infected mothers. It also worked on develop-ing a vaccine against methicillin-resistant *Staphylococcus aureus* (MRSA).

However, the small company had no approved products and zero revenue. Some considered Bancel to be aggrandizing. Yet, for all Moderna's lack of results, its cutting-edge technology kept investors enticed.

INDUSTRIALIZING SYNTHETIC BIOLOGY

The state of synthetic biology back at the beginning of 2020 was a bit like the story of Tantalus, offering but rarely delivering on its promise. The concept of life as programmable was, however, appearing in popular news with the growing meat-alternative industry. The overall need was to lower costs and to increase production. But that was proving hard to accomplish. Amyris was now succeeding in its line of personal-care items but not in biofuels. Bolt Threads, which could make neat synthetic silk ties and hats, at high cost, was looking at mushroom root systems to make leather alternatives. The tool companies like Ginkgo and Twist had inspired large investment. Yet not one of these new companies offered profits. One promising company, a West Coast version of Ginkgo called Zymergen, had an instructive story to tell.

Based in Emeryville, California, Zymergen created a microbe that could make material for foldable cell-phone screens. By 2019, the company had swollen to 800 employees and was ensconced in a new office alongside some half-dozen of the other most prominent synthetic biology companies in the United States. The tall, thoughtful chief science officer and cofounder, Zachary Serber, worked in jeans and a vest, with ear buds hanging around his neck. Growing up near Columbia University in Manhattan, where his father was a physics professor, he was inspired to read widely by a mother who was a high school English teacher. In California, he worked on the Keasling artemisinin project as a doctoral student. "Today we could do the same thing with a tenth or a fiftieth of those resources," he told me. Then he became a scientist at Amyris where he learned the lessons of its tumultuous early years.

Zymergen provided lab services to companies designing pest-resistant plants, plastic-eating microbes, and bio-based fuels. As years passed, however, it was vague in defining its core business, and its main product Hyaline did not progress as predicted.

Across the world, new centers of biological manufacturing sprang up in Israel and Singapore, small countries needing to invest heavily in agriculture and livestock alternatives, and in China and England, while synthetic DNA and RNA were making progress in applications for medicine. In Philadelphia, the Wistar Institute modified DNA to create antibodies for Zika virus and vaccines for the Middle East respiratory syndrome. In Duarte, California, researchers engineered the genetic material of prostate cancer tumors to make them vulnerable to the body's immune cells. In Boston, Massachusetts, Wyss Institute scientists worked on turning cells into tiny programmable factories producing sensors of inflammation.

The efforts to alter DNA and RNA made an extraordinary quest with some singular early achievements. The Venter team showed genomes could be designed, made, and then booted up. Romesberg had built a novel form of DNA, proving "some of Benner's ideas really work," John Glass told me.[33] Other researchers, at the Medical Research Center in Cambridge, England, and at Harvard in the United States, made *E. coli* with synthetic genomes. An international team remodeled the chromosomes of yeast. Others modified plant and mammal DNA. The synthesis of larger and larger genomes enabled pig-to-human organ transplants, offering the prospect of manufacturing genetic medicines and sustainable products.[34] As DNA editing and manufacturing techniques became cheaper, faster, and more powerful, hundreds of labs and start-ups used synthetic biology in creative new ways.

At the same time, the tension between the profit makers and the idealists was growing. Some researchers, inspired by iGEM, DIY biology, and MIT-based BioBricks, favored open-source and immediate sharing of data. That philosophy had been codified in an international agreement in 2014 called the Nagoya Protocol, which offered rules for fair access to the benefits of modified organisms. On the other side was the need to profit from research, as patent protection was established to do. The potential was huge. As many of the companies planned to go public, even the idealists lined up to buy shares.

What could this new world look like, and what role would synthetic biology play in it? "Until we have useful applications," Bolt Threads' Dan Widmaier told me, "we're going to fall into this academic curiosity bucket, rather than 'this is an amazing, world-changing technology.' And I think this is a world-changing technology."

Useful applications make our subject to explore in part II.

II

RIPPLES IN THE WATER

6

RUSH

Biology-Made Medicines

Nature makes drugs, we can too.
—Christina Smolke, Stanford

O n New Year's Eve of 2009 in Pasadena, California, twenty-eight-year-old Stephanie Culler stepped off of the elevator into her professor's darkened lab. Friends said she was crazy to be working. Culler flipped on the lights of the Spalding Lab room to peer through a microscope. Raised in Southern California, Culler lost both of her beloved grandmothers to cancer, one a strong Jewish woman who had survived the Iranian Revolution. "What I wanted in life," Culler recalled, "was to solve cancer."[1]

At Caltech, Culler had planned to study directed evolution with Frances Arnold, but then Arnold herself was diagnosed with breast cancer. Culler thus became the first graduate student to study with a dynamic new assistant professor of chemical engineering, Christina Smolke, pursuing their idea to use cutting-edge RNA synthetic biology to make medicines. The only trouble was that Culler's experiment had not worked in four years, and her advisor was leaving Caltech in four months.

"I was working sixteen hours a day," Culler recalled. "I wasn't eating or sleeping."[2]

Christina Smolke, now W. M. Keck Foundation Faculty Scholar in the Department of Bioengineering and Chemical Engineering of Stanford University, wanted to engineer yeast cells to construct plant-inspired medicines. She was using computational design tools with a goal of changing the way medicine treats pain.[3] Smolke would go on to cofound the company Antheia to produce safer therapeutics, drugs for opiate addiction, and painkillers, with other manufacturers competing from Canada and elsewhere. The shared goal was a more sustainable way of manufacturing pharmaceuticals without plants, by using a combination of biotechnology, gene editing tools, and whole-cell engineering to turn yeast and bacteria into medical factories.

Companies were applying synthetic biology to therapies at a time when antibiotic resistance and drug addiction were public health crises. In an aging world, more patients suffered from cancer, chronic inflammation, and organ failure. To respond, other researchers attempted to program stem cells to regenerate aging tissue or blood platelets. Many sought to program a mainstay of the body's immune cells, called T cells, to attack cancer and other diseases. Companies such as Tierra Biosciences, Kalion, Arsenal-Bio, Demetrix, Synlogic Therapeutics, and Persephone Biosciences explored concepts such as "living medicines," programming the body's natural microbes to detect and treat disease. In England, CHAIN Biotech was modifying one species of the bacteria *Clostridium* to attack colitis.[4] Other companies pursued programmable vaccines from life's genetic code.

None of it was easy. Tierra's young founder, Zachary Sun, admitted, "Every day here is organized chaos."[5] Culler had never worked so hard in her life. "I went home and cried as my way of coping." The efforts included moving the gene pathways normally found in plants into yeast or bacteria or applying the technologies of artificial intelligence to pinpoint each individual patient's genetics, and programming that patient's own immune cells to attack tumors. In Culler's case, the idea was to engineer cells by using the body's messenger, RNA. The immediate goal was to make medicines, but the bigger quest was to make them cheaply, and more, to create new ways of healing the human body.[6]

The medical applications of synthetic biology begin with the immune system, followed by medicines from plants, then stem cells and living medicines, and, finally, DNA-based vaccines. From birth to growth, health to sickness, the story of synthetic biology therapies takes us from small entrepreneurs to the biggest companies, from Galileo's university to the La Jolla Symphony, and we explore some fascinating ideas along the way.

An early idea was to fight cancer.

A STORY OF CAR-T CELLS

In Chicago, Jacelyn Walsh, an athletic, forty-one-year-old mother of two young girls, was exhausted. For six years she had been fighting acute lymphoblastic leukemia (ALL) and wanted to give up. But she could not. "Having children gives you something to fight for," she told her doctors at the University of Chicago Medical Center.[7] Walsh signed up for a clinical trial in which her severely depleted T cells, were removed from her blood so doctors could insert genetic instructions to those T cells to hunt down her specific leukemia cells. T cells are professional killers. As Walsh endured more rounds of chemotherapy, leaving her more exhausted, her own engineered T cells were returned to her bloodstream through an IV catheter.

Engineering the body's T cells to attack tumors had long been a dream of immunologists. Cancer's great danger is its ability to hide from these foot soldiers. One solution involved reengineering the soldiers to detect signature proteins on tumor cell surfaces, and then reinserting them into the patient's bloodstream. The designed killers were named chimeric antigen receptor T cells (CAR-T cells). Antigen receptors are proteins that recognize the surfaces of tumors, viruses, or bacteria. Chimeric, from the ancient Greek word *chimera* (a mythical creature mentioned in chapter 5 that combines parts of different animals), references the fact that the engineered killer cells combined three proteins, one recognizing a cancer cell and the other two signaling the T cell to activate.

It all began in 1989, when Israeli immunologists Zelig Eshhar and Gideon Gross first genetically modified T cells to destroy cancer cells. For thirty years, the process was refined and developed. In the United States, University of Pennsylvania Medicine pioneered the first CAR-T cell medicines.[8] By 2018 two companies, Novartis and Kite, had developed CAR-T cell therapies for the clinic. Walsh received the Novartis therapy, which modified her T cells to attack her tumors. In 2020, after a relapse, she was cancer-free. By 2022, some 600 clinical trials of CAR-T cell therapies were under way, for diseases from myeloma to glioma, leukemia to lymphoma. Companies such as Allogene, Bluebird, Precision, and Shire showed they could engineer T cells to attack different cancers in mice. Other CAR-T cell companies included Arsenal Biosciences in California, JW in China, and GenMoab in Germany. By 2021, pharmaceutical giant AbbVie signed a deal with Jennifer Doudna's gene editing company, Caribou Bioscience, to collaborate on similar therapies.[9]

To be sure, the FDA had approved CAR-T cells for only three cancers, and only for use if other therapies had failed.

The engineered cells had not been effective yet on hard-tissue tumors, which are more difficult to neutralize. The cells were also expensive, hard to manufacture and to administer.

Still, the potential was there. Design of CAR-T cells was one of the fastest-growing applications of synthetic biology to medicine. The approved CAR-T medicines included Kymriah, Yescarta, Breyanzi, Tecartus, and Abecma. Kymriah was for patients up to age twenty-five with acute lymphoblastic leukemia, one of the most common and lethal forms of cancer in young people, as well as for adult patients with follicular or diffuse large B-cell lymphoma.[10] Breyanzi and Yescarta were for adults with large B-cell lymphoma, the most common type of non-Hodgkin's lymphoma, when a first treatment did not work or the cancer returnedin a year. Tecartus was for patients with acute lymphoblastic leukemia. Abecma was for adults with multiple myeloma, a cancer that forms in a white blood cell called a plasma cell, in patients who received four treatment regiments that had not worked.[11]

One of the leaders of the field was Yvonne Chen of UCLA. Another was Wendell Lim of UCSF, who focused on protein switches called synthetic notch receptors. Synthetic notch receptors responded to signals in specific, flexible ways that might expedite the programming of CAR-T cells.[12] "I had no idea it would work," recalled his former doctoral student and University of Southern California professor Leonardo Morsut, a sweet, towering, former Italian national volleyball star.[13] Morsut began his career at the University of Padova, where Galileo taught. "Yeah, sure, Galileo . . . it seemed normal to me," Morsut recalled with a smile. Lim started a company that was bought by Gilead Sciences, and then his team found three new potential CAR-T switches, "on the far end of SynNotch," he told me, ready for Phase I trials.[14]

The bottom line was that CAR-T cells had the potential to translate into the clinic on a wide scale. Allogene's cofounder, Israeli-American; immunologist Arie Belldegrun, estimated the market to be worth more than $13 billion.[15] Modified immune cells could create a sense-and-destroy network out of a patient's own body.

Another area of interest for synthetic biology in medicine was pain medications.

RUSH

When Amyris CEO John Melo learned over dinner that a relative was using cannabidiol (CBD) for back pain, he had an idea. Why not make a painkiller with his company's yeast fermentation processes?[16] Others had the same idea: synthetic cannabinoids, used as painkillers, were bidding to be the next big thing. They could replace the dangerous oxycontin and augment drugs such as CBD. Companies such as Demetrix in Berkeley, Willow in Calgary, Cronos in Winnipeg and Toronto, and others, were producing cannabinoids through yeast fermentation. A company called Canopy Growth and even soft drink giant Pepsi were looking into possibilities of drinks made with synthetic cannabinoids, including coffee that could address chronic pain.

At Stanford, biological engineer Christina Smolke had bigger plans. She turned to making yeast buds into sustainable biomedical factories to replace plant-based medicines. At Smolke's company Antheia, researchers maintained that we too often lack safe, effective drugs because of limitations in obtaining them from plants. Take the opium poppy as an example. This plant contains the compound morphine, our main therapeutic for pain. The problems of extracting morphine from the poppy are daunting. More than 100,000 hectares (think the size of a Caribbean island) of the plant are grown annually. Extracting the compound requires toxic chemicals and money spent to separate morphine from other parts of the plant, all of which results in more than 5.5 billion people having insufficient access to painkillers.[17]

Smolke's team adapted baker's yeast to make compounds found in morphine. This approach involved sequencing the genomes of diverse organisms to combine with the DNA of yeast, which could convert sugars and amino acids into the desired complex molecules. Within the constraints of biology, the yeast can produce compounds through the localization of proteins to different organelles, what Antheia's chief science officer, Kristy Hawkins, calls "whole-cell engineering."[18] Such cell engineering allows Antheia to harness the power of nature on a broad scale, seeking new therapeutics for illnesses such as cancer, HIV, hypertension, and more. The process could be more cost-effective and reliable than that of plant extraction.

Antheia's first molecule—a key starting material for medications—was chemically equivalent to that extracted from plants. In 2018, Smolke's lab moved some twenty-five genes to make enzymes in yeast to produce an opium-based cough suppressant called noscapine, in use since the 1930s, which the company also hoped to develop into an anticancer agent. In June 2021, the company announced that it raised $73 million in series B, or second round, financing as it prepared to bring its first pharmaceutical compound to market. As of July 2021, it had harnessed the biosynthesis of four classes of plant-based medicines for diseases that are currently considered "undruggable."[19]

Antheia had competition. In Winnipeg and Toronto, the synthetic biology company Cronos was pursuing the goal of creating marijuana from genetically engineered yeast. "Cannabis is a complex plant," said CEO Mike Gorenstein. "It features many different genes other than simply those regulating THC production."[20] Cronos signed a deal with Ginkgo Bioworks and was making sustainable chewable gummies called Spinach Feelz at its Winnipeg fermentation plant. Its tetrahydrocannabinol (THC) and CBG gummies were sold in Canada, where the Montreal-based company called Hyasynth was trying to do much the same thing.

Berkeley-based Demetrix pinned its sights on yeast fermentation to produce rarer cannabinoids. The company named itself after the Greek goddess of harvest to "help the world benefit from the world's rarest ingredients," said CEO Jeff Ubersax.[21] Its aim was to treat inflammation, anxiety, and tumors, by developing cannabinoids beyond THC and CBD. To do so, the scientists at Demetrix studied rare molecules in hemp and marijuana plants through a combination of automation, biotechnology, and computing tools. In studying the DNA of the cannabis plant, Demetrix scientists discovered how they could most efficiently combine the DNA of the cannabis plant with that of yeast. Yeast fermentation controls allowed them to make products with higher purity and consistency than those of products of natural sources. The first rare cannabinoid Demetrix planned to take to market was called cannabigerol (CBG), the cannabinoid most others are derived from.

The challenges, however, remained. It was difficult to engineer yeast or bacteria to make compounds they did not normally make, some that may even be toxic to them, and obtain a large output. Sometimes the yeast buds gorged on their nutrient broth and then stopped growing, but in industry, companies want their factories to work continually. One solution to preventing such bottlenecks came from the University of Warwick in England, where researchers inserted feedback loops to extend yeast's protein-making machinery. "It's been hugely exciting," said 's Declan Bates, professor of bioengineering at the University of Warwick, using a phrase that could be applied to synthetic biology as a whole, "to see an engineering idea developed on a computer, work inside living cells."[22]

In research tapping nature's systems, another keen area of interest was stem cells, those embryo workers that can build or revitalize any tissue in the body. Tapping into these tiny fountains of youth had been the dream of scientists since the regenerative powers of these cells were discovered. Several teams and companies set out do so, giving me an opportunity to try being a scientist for a day.

"LET'S GO FLOW"

Tara Deans of the University of Utah was in graduate school when she told her supervisor, synthetic biologist James Collins, then of Boston University, that she wanted to work on mammals, not the simple bacteria or yeast most everyone else was engineering. He gave her three months to see if she could move a complex gene pathway into a bacterium. Her first success was such a breakthrough that, commenting on her achievement, a University of Massachusetts–based competitor marveled, "I've tried all my career to do that. She got it on her first try!"

Now Deans was working on stem cells, the body's builders and rejuvenators. In the beginning of life, the embryo's stem cells divide miraculously into all of the body's varied tissues. Adult stem cells, however, divide only into the same tissue in which they are located. It was long thought that an adult cell could not return to its more powerful immature state, but this was proven incorrect in 2006, when researcher Shinya Yamanaka of Japan's Nara Institute discovered that a small number of genes in mice could restore to an adult stem cell its embryonic superpowers. The prospect of programming such cells to replace aging tissue set off a gold rush of longevity research. But scientists did not fully understand the signals that change adult cells into embryonic stem cells or how to control what organ they might create. Researchers like Deans and others were attempting to better understand the process; in her case, to fight cancer.

Motivated by a close friend's blood cancer diagnosis, Deans earned a National Institutes of Health New Innovator Award to try to program stem cells to generate platelets that, she thought, might prevent cancers from metastasizing. One summer day in 2021, she allowed me to help in her lab in the Utah foothills close to Big Cottonwood Canyon.

Our day at the Sorenson Molecular Biotechnology Building began with a view of the Great Salt Lake. The Deans lab overlooked a grass field in the shape of the state of Utah, with mountains in the background. She wore black jeans and a sweater and explained I would be checking whether her team had successfully moved genetic parts of the stem cells' on-off switch from a more complex cell into a bacterium. A mother of three, Deans usually got started early in the morning. If she got in early enough, she could watch a family of deer in the field. We walked past the cell culture room where her students did their stem cell engineering. "Bacteria are not allowed," she joked.[23]

Her undergraduate Travis Seamons patiently walked me through the process of feeding a machine called a flow cytometer to confirm whether their inserted gene was working in their *E. coli*. Their hope was the inserted gene could produce proteins for therapeutics. Trimly bearded and dressed in an open blue shirt over a white T-shirt, blue plastic gloves, and tan jeans and sneakers, Seamons has explained this work previously to Utah's state representatives. At the whiteboard, he diagrammed the restriction enzymes that cut the DNA and allow new circuitry to be inserted. While he spoke, bacteria with his engineered DNA shook rhythmically in plastic tubes in an incubator. Seamons cheered me on: "Alright. Let's go flow."

Downstairs, in a narrow, black-curtained room, I gingerly inserted the test tube containing the engineered bacteria into the sleek white cytometer, which resembled a loud dentist's machine. The cytometer shines lasers through bacteria, running the cells in single file to measure how the light was diffracted, showing us their size and the fluorescence of the inserted gene. If the gene circuit was working, the cells would glow green. "The brighter they shine, the stronger the circuit," Seamons said. The machine vibrated and blinked. Seamons and I watched the computer screen. The cells lit up. Their circuit was working in living cells. The first time it happened, Seamons recollected for me, "I thought, no one in the world has ever seen this before!"

In her project, Deans studied how to enhance stem cell differentiation for the purpose of mass-producing platelets and red blood cells to improve blood supplies in hospitals. She was also engineering stem cells to be delivery vehicles targeting cancer. She hoped this would help prevent tumor cells from spreading. Leading companies developing stem cell therapies included bit.bio, Celularity, ViaCyte, and Rubius Therapeutics. Many others, however, marketed unregulated stem cell therapies, and the FDA has warned consumers against falling for promises of restored youth. As of this writing, the only FDA-approved therapeutic use of stem cells consists of blood-forming stem cells derived from cord blood.[24]

Still, several labs were using stem cells to grow tiny versions of human organs called organoids. In 2021, a university hospital lab in Dusseldorf, Germany, grew mini-brains that developed, on their own, rudimentary eye structures.[25] If adult stem cells could be reprogrammed into their embryonic state, they might be used to heal spinal cord injuries or neuronal illnesses like Lou Gehrig's disease, help stroke victims, reverse retinal degeneration or muscle damage in multiple sclerosis, or, even, fix hearts. With these advances, some of the biggest health-care companies, like Johnson & Johnson and Pfizer, were starting their own projects in stem cells. The path to clinical use remained a challenge riddled with big claims and disappointing findings, where some promising avenues failed or proved prohibitively expensive.

Another important idea was to engineer the body's gut microbes to detect and attack disease. The term for that was "living medicines," and it was the goal of several new companies, including one run by Stephanie Culler and another, Synlogic Therapeutics in Cambridge, Massachusetts, led by another young woman, Aoife Brennan.

THE DESIGN OF LIFE

When the imaginative, poetry-writing endocrinologist Aoife Brennan was in high school in Ireland, she tended her parents' small rural bar south of Kilkenny. "I learned everything I needed to run a biotech company by working at a pub," she told me. "It made me resilient, dealing with difficult people." She laughed. "I learned how to turn a profit, and how to listen."[26]

Brennan, a graduate of Trinity College Dublin and Harvard Medical School, had led Synlogic since 2018. At Harvard, she worked in translational medicine, the art of taking discoveries from the lab to the clinic. Afterward she worked at several gene therapy and gene editing companies, including five years at

Biogen. Then a recruiter called her about becoming chief medical officer at Synlogic, and she became fascinated by engineering the normal human gut microbes to detect early-stage disease. Soon she was applying her pub-tending to running the small biotech company, partly from her home during the pandemic, and with MIT's Jim Collins as a scientific advisory board member. "She is the stubborn Irish woman who won't give up," he said of Brennan.

Synlogic worked on the principle that the body is an ecosystem, and if you can train the ecosystem to monitor itself, you can save lives. Its aim was to produce more effective drug treatments tailored to meet patients' specific needs by transforming some of the body's normal microbes into synthetic biotic medicines. As of 2022, Synlogic is involved in clinical development after successful trials of treatments for metabolic disease and for cancer.

The company signed a deal with Ginkgo Bioworks as its investigative platform, a partnership that paid off with the development of SYNB1353, an engineered microbe for treating a disabling disease called homocystinuria, which is caused by excess methionine, an amino acid, in the blood. The disease results in bone defects, blood vessel obstruction, and intellectual disability, which the Synlogic microbe averts by consuming the excess methionine in the gut. After filing an investigational new drug application (IND) for the medicine with the FDA, Synlogic announced Phase 1 clinical trials had shown proof of mechanism for the treatment.[27]

Then there was Persephone Biosciences in San Diego, cofounded by Stephanie Culler to develop the potential of the gut microbiome in precision medicine, in particular for cancers such as the ones that killed both of her grandmothers, one dying from colon cancer and the other from lung cancer. Culler was a driven researcher who played concert violin with the La Jolla Symphony.[28] One night shortly after the New Year's Eve failure of 2009 when still a graduate student, she was in the lab again at 2:00 a.m. to peer through the microscope. This time, to her amazement, the human gene was glowing with an eerie green light. She had programmed a human cell to turn on a gene that could detect cancer. If it detected proteins made by tumor cells, it would turn on a death gene, killing itself. The technique was modular, rapid, and deployable to detect "a wide range of cancers," she said. After five years, she had done it. She raced to write up the result, which appeared in the journal *Science* and was highlighted as one of the key breakthroughs of 2010.

Determined to attack cancer, Culler felt she needed industry experience and began by engineering microbes at San Diego company Genomatica to make chemicals traditionally made from petroleum. There, she was part of the research and development team, first to commercialize microbial fermentation of a major chemical used in plastics, an achievement that won a Presidential Green Chemistry Challenge Award for Genomatica. Like Brennan, she was becoming fascinated by the microbiome's effect on health. When her grandmothers had received the same cancer therapies that were successful for other patients, for them and others the therapies had not worked. The fault, a series of papers were suggesting, lay with the microbes of the gut. All disease, Culler liked to quote Hippocrates, begins in the gut, which comprises some 80 percent of the human immune system. If you transferred the gut microbes of metastatic melanoma patients into the gut of mice with cancer, the mice responded to the immune checkpoint inhibitor drugs that had not worked on her grandmothers.[29]

In 2017, Culler cofounded Persephone Biosciences to explore ways of using synthetic biology to restore the normal natural gut microbiome in cancer patients. By the fall of 2021, Persephone signed a deal with Janssen Biotech to study the microbiomes of cancer patients who had responded to cancer therapies. It was licensing its engineered microbiome therapies for patients with COVID-19. With grants in hand, the company launched an ambitious program called Poop for the Cure, to study the gut microbiomes of dogs and humans, both cancer-stricken and healthy. It did so by collecting stool samples from around the country. In 2022, it launched an infant version of that study to collect baby stool samples in an effort to combat food allergies.

Engineering microbes to make pharmaceuticals and other chemicals was a part of the work at another Massachusetts start-up, Kalion, co-founded by Kristala Prather, Arthur D. Little Professor of Chemical Engineering at MIT, and her husband. Prather was one of the early SynBERC members whose career at MIT featured breakthroughs in getting microbes to make glucaric acid, a key ingredient in the drug Adderall, but also a sustainable chemical to improve water treatments and detergents. Growing up as a Black girl in small-town Longview, Texas, Prather loved tinkering around the house. After her mom dropped a necklace down the drain, Prather told her, "Give me a wrench," and recovered it. Discussing college applications with her high school history teacher, she said, "I like math and science."

"Put those together," the history teacher said. "That's engineering. You should study chemical engineering and apply to MIT."

She did not know what MIT was, but applied and was accepted. There, Prather told me, "I loved the idea of cutting and pasting pieces of DNA, combining biochemical engineering and synthetic biology." After graduating in 1994, Prather earned a

PhD from the University of California, Berkeley, working in the Keasling lab and then, years later after she had become an MIT professor, participating in the critical SynBERC meetings. Like Culler, she felt she needed industry experience and worked at Merck Research Labs for four years before returning to MIT as a professor and eventually cofounding Kalion. Prather's lab was modifying DNA to be added to microbial cells. "We want to turn them into factories," she said, making biomass-derived chemicals to replace fossil fuel–based chemicals.[30] The young girl who loved to tinker with objects was now tinkering with the design of life.

Synlogic, Antheia, Cronos, Persephone, and similar start-ups were manipulating genes to produce microbes for therapeutic or manufacturing functions. The field was advancing fast. Synlogic presented proof of mechanism in humans from two trials, and the industry site STATNEWS declared it one of the ten companies in synthetic biology to watch in 2020, eager to begin clinical trials for a study in solid tumors and lymphomas. Persephone signed a deal to collect microbiomes from patients who responded well to cancer drugs. Kalion partnered with a manufacturer to expand its production of glucaric acid, identified by the U.S. Department of Energy as a "Top Valued-Added Chemical," and other industrial chemicals.[31]

From CAR-T cells to stem cells, from the microbiome to the microbes making molecules for medicines and products, synthetic biology companies were attracting more and more money on the cusp of the SARS-CoV-2 outbreak. The pandemic would put synthetic biology further into the spotlight.

Before all that, two standbys of academia and medicine got involved in modifying DNA to make vaccines, in ways somewhat similar to what Moderna and BioNTech were doing with RNA. At first, few people noticed.

OUT OF AFRICA

On the other side of the Atlantic, Oxford-based vaccine researcher Sarah Gilbert and her team pursued the idea of using DNA to make vaccines. She and Adrian Hill, director of the Jenner Institute, were modifying a cold-like virus found in chimpanzees to prevent patients from contracting Ebola. The method was like that of traditional vaccines: make a harmless version of a chimpanzee cold virus, get it into human cells, and allow the immune system to develop its defenses to attack it. The red-haired Gilbert was a pragmatic, hard-edged researcher. The mother of triplet boys relaxed, infrequently, by playing her saxophone in the woods so as not to disturb the neighbors.[32] Describing her routine, Gilbert woke up most mornings at around four o'clock "with lots of questions in my head," she told Bloomberg News.[33]

Sarah Gilbert and Adrian Hill were a dedicated team on a social mission. Hill had been inspired by visiting his missionary uncle fighting Ebola outbreaks during the Zimbabwe civil wars in the 1980s. Blond and rumpled in his tweed brown jacket, the childlike sixty-one-year-old Hill planned clinical trials of vaccines for Ebola Virus Disease and for Middle East respiratory syndrome (MERS-CoV). MERS was a disease that had been a scourge in countries such as Egypt. Originating in bats and transferred to camels which then infected humans, MERS was a harbinger of the coming COVID pandemic.

Using gene editing techniques, the Oxford researchers modified nucleic acids with instructions to make a protein from the virus surface. The patient's body developed a tuned immune response to the virus protein. By early 2020, the first company they co-founded, Vaccitech, was finalizing clinical trials for its Ebola and MERS vaccines.[34] This preparatory work would prove critical when the COVID-19 pandemic broke out.

The Oxford technology did not boggle minds as much as wonderkids mRNA and CRISPR might, but it did not have to be frozen and could be given in one dose, not two. Even as SARS-CoV-2 spread, Oxford continued work on its DNA platform for Ebola. By November 2021, the University of Oxford was recruiting for clinical trials of its vaccine. In another part of the world but toiling on a similar platform, a giant company known for painkillers, Johnson & Johnson, had a similarly straightforward synthetic biology idea for vaccines.

JOHNSON & JOHNSON

From Band-Aids to baby powder, market-dominating Johnson & Johnson was one of the world's largest and most well-established health-care companies. Since its founding in 1886, the $394 billion multinational health-care empire, with more than 250 subsidiaries across 175 countries, was at the top of Big Pharma's list of powerful corporations. While the company was best known for its Tylenol and first aid supplies, it featured three main sectors: consumer health products, medical devices, and pharmaceutical products. With Tylenol as one of its staples, Johnson & Johnson was a maker of drug and first aid products present in most every household. But when it came to vaccines, it was a relatively small player. Until 2020.

Twenty years earlier, Harvard vaccinologist Dan Barouch wanted to create a vaccine for HIV.[35] He was focusing on adenovirus serotype 26, called Ad26 for short, a fairly rare human virus that causes mild colds. Barouch tried treating animals with an Ad26 virus containing an HIV gene, with the goal of arming the body's immune system against HIV. While trials were promising in monkeys, no human trials had been conducted.

Johnson & Johnson was interested, and Barouch became affiliated with the company. As the Zika epidemic emerged in 2015, his team prepared a vaccine using Ad26 and Zika spike proteins, only to shelve the vaccine after Zika retreated.

By the beginning of 2020, climate change and human overpopulation were contributing to an increase in zoonotic viruses, those that passed from animal hosts to humans. The flu-like viruses included the severe acute respiratory syndrome (SARS) virus in China in 2003 and the MERS virus in 2011 in Egypt, Saudi Arabia, and beyond. Zoonotic viruses also included Zika and Ebola, all deadly enough, but still the lull before the SARS-CoV-2 storm.

In that lull, the pharmaceutical side of synthetic biology was laying groundwork. Multiple labs and companies were in pursuit of synthetically engineered CAR-T and stem cells, non-addictive painkillers, and living medicines for diseases ranging from depression to high cholesterol. Therapies based on modified CAR-T cells moved to clinical trials, totaling 836 in planning or practice in 2021.[36] Adult stem cell clinical trials in 2020 numbered some 3,000.[37]

At the opening of 2020 synthetic biology's successes—in malaria, fragrances and cosmetics, and meat alternatives—had still been few. However, propelled by ecological cataclysms such as raging fires, rising sea levels, and a floating island of plastic waste, one pressing avenue of research was the environment, which is where we turn next.

7

NEW NATURE

A Do-It-Yourself Environment

Sometimes we picture it as an echo of the original Garden of
Eden . . . there is nothing fainthearted or wimpy about plants.
—Diane Ackerman, *The Rarest of the Rare*

Dragonflies buzz and mosquitoes bite on South Florida's
Gulf shore at dusk. Red mangroves sit on alien roots,
their salty arms stretching to the darkening sky. Motoring out slowly past the coastal groves, beneath their twisted,
gothic branches, I imagine the first Spanish ship as viewed by
the Calusa Indians and all that meant for the environment.

On the eve of the pandemic, several researchers were applying synthetic biology to the environment. Colorado State University's June Medford offered an idea for desalinating water the
way mangroves do. Northwestern University's Julius Lucks and
others edited genes to detect and clean polluted water. Others
sought to use gene drives to attack diseases by spreading dominant genes into the wild to control populations of pathogen-carrying insects and animals. The University of Guelph's Rebecca
Shapiro was modifying such a gene drive to kill off a drug-resistant fungus found often in hospitals.[1] Some researchers were

editing the genes of coral so they might resist warming oceans, while others tried to create hardier plants and crops, natural fertilizers, and green chemicals—all without petroleum.

By 2020, at least 3 percent, or $400 billion, or perhaps more, of the U.S. economy was based on biotechnology.[2] But synthetic biology to heal the environment kept hitting snags. Bioplastics did not perform well at first, challenging the giant Danish toymaker Lego, which committed to a fully bioplastic product line by 2030.[3] So many soft drink companies faced problems that Coca-Cola offered to share its secret bioplastic formula with some of its competitors. Still, Danimer Scientific was expanding production of bioplastic for straws for Starbucks and Dunkin' Donuts at a Winchester, Kentucky, plant.[4]

Around the world, efforts were continuing to make microbes that degrade plastic waste, or cool the Earth, or generate electricity, or help crops require fewer pesticides and less water and fertilizer. Others sought to protect people from Lyme disease or West Nile virus, ensure the health of wild animals, kill off the mosquitoes that carried malaria, yellow fever, and Zika, and detect and safeguard against other deadly pathogens. At the University of São Paulo, plant physiologist Lazaro Peres was restoring the natural genetic diversity of crop tomatoes.[5] As of 2021, gene-edited plants were being grown by 17 million farmers in twenty-nine countries.[6]

The motive was to repair a world where humans were causing problems such as an extinction crisis, an alarming increase in zoonotic diseases like Ebola and COVID-19, non-native species invasions, growing waste and contamination, drought, and floods. To confront climate change and environmental encroachment, several synthetic biology efforts were under way, including altering the microbes in the guts of cattle, whose releases of methane had a major warming effect, or engineering pine trees to be more fire resistant. A race to restore cooling grassland to the Arctic tundra produced an unlikely union of Siberian and Harvard scientists in a remote preserve they called Pleistocene Park. The Florida panther, inbred and near extinction in the 1990s, had been revived through introducing its cousin, the Texas puma, into its habitat. Scientists proposed restoring the virtually extinct American chestnut tree by introducing a gene from wheat that could help it resist the fungus that killed.[7]

Some researchers proposed releasing genes into the wild to eradicate invasive New Zealand and Australian rodents that decimated native birds. Others wanted to bring back extinct animals through engineered DNA, *Jurassic Park* style. The San Diego Zoo maintained a "frozen zoo" of DNA from endangered or recently extinct species.[8] A California-based nonprofit organization, Revive and Restore, went a step further with a new science called "genetic rescue" to maintain threatened or extinct species through a variety of cell biology tools.

It might seem oxymoronic that one could preserve nature with synthetic biology. To understand how researchers were redesigning organisms for environmental benefits, we journey to meet doctors in West Africa, iguana hunters in South Florida, coffee growers in South America, and corn farmers in Iowa. We begin with mangroves.

TO THE SEA, THE SEA

The wind is whipping the Gulf of Mexico as I head to the Ten Thousand Islands, the barrier islands south of the Florida Everglades. Ahead, the Gulf shimmers in the last pink rays of sunshine, home to giant grouper, sharks, marlin, oil rigs, smugglers, and mangroves. Bordering this fragile ecosystem are mile upon mile of tangled trees that protect the coastline from hurricanes, provide homes to numerous species, and even remove salt from salt water.

The mangrove is a salt-loving plant found at the boundary of land and sea in the tropics and subtropics. Along the coastlines and islands of Florida, the Caribbean, and South Asia, these trees form intricate underwater root systems that serve as havens for fish, sea birds, alligators, sharks, and turtles to lay their eggs and also provide critical protection against hurricane surges. Mangroves desalinate salt water and filter farm runoff. Green-leafed and hardy, they create mazelike, dense canopies that stretch over hundreds of miles. With their tangled roots resembling the landing gear of Martian rovers, they take in salt water and seal in precious freshwater while secreting salt through the surface of their leaves, hence their scientific name, halophytes.

Mangrove trees are also great at computation, Colorado State University's June Medford tells her classes. They use energy from a food source 93 million miles away, with little more than salt water and carbon dioxide for sustenance. But what if these haunted, sheltering trees could purify water for the thirsty billions of human beings on Earth or irrigate the crops that feed them?

That was what June Medford sought to do. After graduating from the University of Maryland, Medford obtained her PhD in biology at Yale. In 2003, she met Nobel winner Frances Arnold at a DARPA meeting and became entranced by the idea of using synthetic biology to edit nature. Winning Defense Department funding, she engineered plants to be detectors of explosives and contaminants.[9] Then she turned to mangroves.

Studying the mangrove, she wanted to recreate the tree's DNA circuits in another plant so that it could produce fresh-water from sea water. To transfer the mangroves' ability to desalinate seawater, Medford tapped a plant virus to deliver three desired genes. Using a technique tested in the biofuel crop prairie switchgrass, her lab created a switch to turn on a metabolic pathway. "The switches were not perfect!" Medford told conference audiences. "But we put enough of the pieces together to prove the concept."[10] Her lab developed an artificial desalination system in the flowering plant Arabidopsis.[11]

One third of the world's irrigable land is contaminated with salt, and the amount of land succumbing to rising sea levels is growing every year. Freshwater is scarce for 4 billion people, and the global crop loss due to rising salt levels in soil totals more than 20 billion pounds of material a year.[12] Inspired by NASA's famous Christmas Day Earthrise photo, Medford believed synthetic biology "could change global perception" in a similar way.

The process was not simple. Their engineered plants at first grew slowly, yielding only small amounts of freshwater. Mangroves use a waxy waterproofing substance called suberin in their roots as a barrier to salt. Medford's team programmed the plants to produce suberin, a mutation that had never been done. The act of plugging in genes at first stunted the plants' growth. As her team refined their gene circuit, the engineered plants began secreting more freshwater comparable in quality to bottled water. By 2021, Medford cofounded a company, PlantMade-Works, to develop her technology.

Several other labs were looking at different ways of making agricultural crops tolerant of salt. A Canadian company, Agrisea, created salt-resistant rice that could be grown in the ocean and was testing it off of the coasts of Singapore and Grand Bahama Island. Their crops used a seagrass form of salt tolerance. Scotland-based Seawater Solutions was seeding flooded coastal farmland with salt-tolerant herbs like samphire and sea blite, creating wetland ecosystems without pesticides. Other companies in Israel, China, the United Arab Emirates, India, and the Netherlands were making salt-tolerant crops or engineering cyanobacteria and microalgae to desalinate seawater. A Yale researcher created an artificial mangrove in the lab.[13]

As synthetic biology was modifying sea plants, another threat, one close to home, introduced the synthetic biology interventions called gene drives.

GENE DRIVES AND LYME DISEASE

Out hiking, a young woman felt something on her thigh. She spotted a small brown tick. Wrenching it off, she worried her life was changed forever.

She was correct to worry. Some 476,000 Americans suffer from tick-borne Lyme disease infections annually.[14] Carried in a microbe in ticks who feed on forest deer and mice, it is so widespread in Cape Cod and Nantucket Island that some people have become desperate. One proposed solution, offered by MIT's Kevin Esvelt in tense community meetings on Nantucket Island, was to unleash genetically modified white-footed mice that resisted the tick-borne pathogen. As islanders met to hear Esvelt pitch the gene drive solution, the proposal seemed enticing.

A gene drive is a gene that spreads itself faster than it normally would in the wild. More precisely, it is a "selfish genetic element that biases inheritance in its favor," explains University of Guelph researcher Rebecca Shapiro. Gene drives are naturally occurring and observed since the nineteenth century, studied in yeast and mosquitoes in the 1960s.[15] In the years since, researchers have pursued ways to manipulate them to control invasive species and protect human health—in mosquitoes to defeat malaria, dengue, and Zika virus, in fruit flies, and in the pathogenic fungi found in hospitals. By 2020, gene drives had been proposed to limit the populations of disease-carrying mosquitoes, New Zealand possums, tick-ridden mice, and other unwanted animals. Enticing, powerful, and inexpensive to deliver, gene drives presented a scientific gift perhaps too good to be trusted.

On Nantucket, Lyme disease infected some 40 percent of inhabitants. Thus, despite the objections, in a series of public hearings islanders signaled they were ready to try a gene drive release. Esvelt received permission to release transgenic white-footed mice in a program called Mice Against Ticks. The plan seemed promising.[16]

The angular, soft-spoken Esvelt had trained in the lab of David Liu, a Harvard collaborator of George Church. At MIT, Esvelt directed the Sculpting Evolution group, inventing new ways to influence ecosystem evolution, seeking to address spreading ecological problems through synthetic biology. Esvelt and Liu helped pioneer phage-assisted continuous evolution (PACE), a microbial ecosystem using the viruses that infect bacteria to speed up evolution in the lab, and Esvelt was an early advocate of using gene drives to alter the traits of tick-carrying white-footed mouse populations.

To be sure, the release of gene drives faced hard questions. The technology was in the early research phase and lacked rigorous regulatory review or extensive field testing. The worry was that such interventions could spread genetic traits beyond seed zones with unanticipated side effects. Some of the most outspoken critics of gene drives included other scholars. "There's a sense that these are things humans are not supposed to be doing," University of Wisconsin law professor Alta Charo told the *New York Times*.[17] To address these challenges, MIT researcher Maud Quinzin and Esvelt devised a technology called "daisy drives" featuring an on-off switch limiting the number of generations a gene drive would persist by dropping pieces of DNA with each new generation, like a daisy losing its petals.

Others tried gene drive experiments in more controlled environments. Trained in the lab of Harvard's George Church also and at the University of Toronto, Rebecca Shapiro had figured out how to insert a gene drive into a single-celled fungus, *Candida albicans*, and showed how quickly the gene could be spread through a population. *Candida albicans* is found in the intestine of up to 60 percent of humans and is the most common infection seen on medical devices such as hospital catheters. A dangerous relative of yeast, it appears capable of crossing the blood-brain barrier and can turn deadly in compromised individuals, like the elderly, AIDS patients, and cancer survivors. Easy to grow, the feared hospital infection has become more and more resistant to standard antibiotics.

Shapiro regarded gene drives as a way to defend against the killer fungus. In a postdoctoral position with Jim Collins at MIT, she developed a *Candida*-killing gene drive system with near 100 percent effectiveness.[18] She bred a mutation to switch on two deadly genes and bequeath those mutations to offspring. The first proof of principle was turning all the microbes fluorescent red in a Petri dish. "It works insanely efficiently," says Shapiro, now a professor at the University of Guelph in Canada, where she is expanding the technique's range.[19]

Another concern raised about using gene drives was their potential use in biowarfare. As a reported in the *Bulletin of the Atomic Scientists*, researchers' publications may reveal information to those who wish to do harm. For these reasons, and with the larger concern about releasing genetically modified organisms into the wild, a 2017 *Science* paper laid out key principles for responsible gene drive research.[20] Some civic organizations, such as the Friends of the Earth Network, proposed a moratorium on gene drives in the wild until further study could be conducted. In 2021, a World Health Organization Guidance offered rules for testing gene drive modified mosquitoes, including the requirement to involve local communities in decision-making.[21]

In West Africa, Brazil, the United States, and other regions, two different efforts to research the modification of disease-carrying mosquitoes and other pests were being run by the non-profit research consortium Target Malaria and a company called Oxitec.

MALARIA AGAIN

"I cannot tell you how many times I've had malaria!" Dr. Mamadou Coulibaly, in a gray dashiki, tells me from his lab in Bomaka, Mali, where he directs the Target Malaria effort at the Malaria Research and Training Center to genetically alter mosquitoes to prevent the disease.[22] Resistant forms of malaria were again spreading, to the extent that a child in Africa died once every minute from the illness. Coulibaly recalled his childhood in the village of Bonankaro, where his grandmother boiled roots for a bath and drink potion when he became ill. The severe form of the disease attacked the brain, causing delirium, convulsions, and death, mostly in children.[23] With more than 600,000 Africans dying every year, Target Malaria was planning to implement a three-step program to target disease-carrying pregnant female mosquitoes, working with public institutions in Burkina Faso, Mali, Ghana and Uganda.

In phase one, Target Malaria engineered a non-gene-drive sterile male mosquito. This strain was imported into Burkina Faso in 2016, studied for two years in the lab, and then released in 2019 to understand what happened when lab-reared mosquitoes were released in the wild. Some community members felt they were not properly consulted at the time, however, igniting one protest in the capital city of Ouagadougou in the summer of 2018.[24]

Phase two was to engineer a fertile strain of non-gene drive "male-bias" mosquitoes, modified to produce mostly male offspring. To create it, researchers in London and Terni, Italy injected mosquito eggs with engineered DNA, tagged with a gene for a fluorescent protein. If the phase two trial runs are successful, the final phase would be the release of gene drive mosquitoes with modifications to make an entire mosquito population infertile. For the moment however, gene drive mosquitoes are created and bred only in Target Malaria's two European labs.

Synthetic biology was being used in other ways to attack malaria, notably in a messenger RNA vaccine in development from BioNTech and another vaccine from Oxford. But these 2022 breakthroughs remained a few years off in deployment. "The tools we have now are not enough," Target Malaria researcher Alekos Simoni told me. "We are working hard to test the technology and will release it only if it is safe, efficient and accepted by the affected communities."[25]

Another form of genetic modification, also eliminating disease-carrying female mosquitoes, was the focus of another company, UK-based Oxitec.

FLORIDA AND BRAZIL

Walking door to door in the Florida Keys before a county referendum Meredith Fensom, Oxitec's head of global public affairs, was listening to residents. The Florida native faced a small shack on a dirt road. A man sat with a gun beside a cooler full of dead iguanas. "I like iguanas, but they're invasive," Fensom told me. He listened to her pitch and invited her to hunt with him. Her father, Fensom recalled, took her duck hunting as she was growing up. "Hunters are conservationists," she said. "They notice changes in the environment more than anyone." The referendum passed.

Oxitec is a U.S.-owned company based in the United Kingdom, with research labs a few miles from the Oxford campus where it was founded. Oxitec's 2021 Florida release of more than 20 million gene-modified mosquitoes made for a test run of a technology to combat an invasive insect that carried diseases harmful to both humans and pets. Brought to America on slave ships, the mosquito *Aedes aegypti* harbored pathogens that caused dengue, yellow fever, and Zika. Florida Keys health officials were interested because of a 2012 outbreak of dengue there. Oxitec engineered a self-limiting gene mutation to reduce invasive mosquito populations.[26] The gene drive is spread by leaving boxes of mosquito larvae in hard-to-find places. Some objected. "Genetically engineered organisms are not something you can control," one opponent told NBC News. Still, the Keys release was deemed so successful that the EPA approved the release of 2.4 million more gene-edited mosquitoes in California in 2022.[27]

A developer of "living pesticides" to prevent human disease, livestock illness, and crop damage, Oxitec modified *Aedes* without a gene drive. Instead, a lethal tTAV gene in the mosquitoes is turned on, killing off female offspring before they can reproduce.

The gene spreads normally in the environment, such that Oxitec's modification disappears in about ten generations. The edit can be switched off by the antibiotic tetracycline. Oxitec claimed the technology could protect Florida residents against four viral diseases: Zika, yellow fever, chikungunya fever, and dengue, two of which had no vaccine. It was so popular in Brazil that more than a billion edited mosquitoes have been released, and the kits are available to farmers for purchase. The argument is that the gene modification is inexpensive and precisely targeted, more selective and potentially safer than mass spraying with harmful insecticides, which kills beneficial insects like bees.

As explained to me by Nathan Rose, Oxitec's head of malaria programs, also released in Brazil was a modified gene to protect the country's corn crop from a killer moth called the fall armyworm. "It's a similar process," Rose explained. "We modify the tTAV gene that kills the female offspring, with a fluorescent marker to tell if the caterpillars come from one of our male moths. With the moths, we release the adults, not the larvae."[28]

Elsewhere in the world, synthetic biology was being used to recycle waste, linking the science with an idea called the circular economy, in which one industry's refuse becomes another's raw material. Around the world, several small start-ups were pursuing that framework, also called closed loop, as a way to remediate Earth.

THE CIRCULAR ECONOMY

After coffee prices skyrocketed around 2016, Denmark-based researcher Juan Pablo Medina, founder of Kaffe Bueno, began buying spent coffee grounds and recycling them into chemicals for cosmetics and foods. Winner of The Circulars prize, the soft-spoken Medina pointed out that only 1 percent of coffee bean compounds is used in making coffee, and that he could recycle the other 99 percent discarded as waste, making his company a part of the circular economy. The circular economy aimed for the elimination of waste through smart product design and complete recycling.[29] In it, one company's waste becomes another's raw material in a practice known as upcycling. Around the world, local entrepreneurs bought into the idea. More than 180 composting and education-based nonprofits and businesses joined together to form an international Upcycled Food Association to share expertise.

A few such companies addressed the issue of environmental classism, in which rich countries dictate to small or poor farmers what to grow and how. One effort in Mexico City sought to help such farmers make more money from their waste: the company Xilinat bought waste from Mexican farmers to convert into sugars in a syrup for sale to the food and beverage industry. Those profits provided needed money to family farmers to educate their children.

On a wider scale, the Milwaukee-based company Agricycle Global modified microbes to regenerate soil and resolve or address food spoilage. Some 2.8 trillion pounds of food spoils every year, as many farmers in developing countries lack refrigeration facilities to preserve produce if prices plummet or items do not sell.[30] Agricycle's field officers buy the unused parts of fruits like peaches and mangoes, almost 95 percent of which is pits, skin, and unused flesh, and grind the pits to make gluten-free, nutrition-rich flour or recycle the peels for extracts that make ingredients for cosmetics or natural mosquito repellents. The company distributed cheap solar dehydrators to preserve unused produce for sale later, and encouraged its network of 40,000 farmers to recycle coconut shells to be converted into charcoal.

Another tack was to prevent food products from spoiling. California-based Apeel Sciences used gene editing to create an extra layer of natural coating for fruits and vegetables to extend their shelf life. The company was one of the top five money raisers in the industry in the first half of 2020, joining fast-growing giants like Impossible Foods.[31]

Meanwhile, several synthetic biology–based programs sought to address climate change and the human factors contributing to it. One of the most unusual projects was an attempt to resurrect the extinct mammoth, a once-dominating animal that last thrived on the planet about 10,000 years ago. Welcome to Pleistocene Park.

ON THE STEPPES

The wind is whipping over a lonely, rusting, Siberian outpost, most of whose population left after the collapse of Communism. But inside the well-lit and warm Quonset hut, American microbrewed beer awaits the hearty few scientists who show up. To get here and witness the unlikely union of Siberian father-and-son adventurers and thinkers, you need only fly across thirteen time zones, the last in a Soviet-era bomber. Pleistocene Park is the much-promoted brainchild of Russian scientist Sergey Zimov, with the goal of preserving the melting Siberian Arctic tundra. These cold upper reaches of northern Russia are rich repositories of mammoth DNA, found in tusks, skin, and skeletons. Zimov theorized that mammoths once helped to preserve the widespread Ice Age grasslands by spreading the grass seeds in their dung and by destroying the trees that were replacing the grass. The grasslands, Zimov and others argued, made for much cooler habitats than forest trees, reflecting back the sun's heat instead of absorbing it. If you restore grasslands, you can cool Earth and ease the biggest problem facing the tundra today: melting permafrost that was reviving previously frozen methane-producing soil microbes. Methane produced one hundred times more warming effect per ton than did carbon dioxide.

The mammoth restoration project has been promoted by George Church in his book *Regenesis* describing his effort to restore the giants' DNA, *Jurassic Park* style.[32] The idea went like this. You implant the mammoth genes into African elephant DNA, nurture embryos in a Petri dish, and possibly implant the embryo into an elephant to grow. The only issue is that the gestation of a woolly mammoth takes two years, and it would be highly challenging either to induce an elephant to give birth or to artificially mimic a mammoth's womb. Lack of money was another problem. So was the fact claims about the cooling effect of Pleistocene-type grasslands had never been tested in a controlled study published in a peer-reviewed journal. Finally, many critics questioned the ethic of creating a single chimeric creature from the past with no companions or ecosystem to support it.

One step on the way to a green synthetic biology revolution was much more American: to process nitrogen normally provided by expensive petroleum-based fertilizer. To achieve that, the company Pivot Bio provided the fertilizer for the lush corn made famous in Major League Baseball's first Field of Dreams game in Iowa in August 2021.

FIELDS OF DREAMS

In graduate school in Iowa City, lab partners Karsten Temme and Alvin Tamsir shared a vision of replacing petroleum-based fertilizer. CEO Temme, described by a former professor as out-of-the-box and quirky, explained, "It takes four things to grow a plant: sun, water, carbon dioxide, and nitrogen. A plant is really good at making the first three, but can't make its own nitrogen." That comes from fertilizer, which is often made from petroleum. Although fertilizer increases productivity, its manufacturing and use accounts for about 2 percent of the world's greenhouse emissions. In addition, half of the nitrogen in fertilizer never reaches crops, but instead ends up in waterways that feed into lakes and oceans, causing dead zones and oxygen-depleting blooms as algae feed on it, killing fish. Increasing the mobilization of nitrogen to plants could reduce emissions for an impact on par with planting 16 billion trees, Pivot Bio's founders claimed.

Founded in 2011, the Berkeley-based synthetic biology firm launched its first commercial product in 2019. Pivot Bio Proven, for corn, sold out within six weeks of its availability.[33] Plants use microbes to help pull nitrogen from the soil. Having sequenced the DNA of nitrogen-fixing microbes from plant roots, the company made a more efficient version of those microbes to attach to crop roots. The reprogrammed microbes sense nutrient needs of a crop and produce the perfect levels of nitrogen through all types of weather. Applied during planting, the Proven microbes and crop seeds mix in the ground. Once the microbes become attached to the root structure, they improve the plants' ability to capture nitrogen, convert it to ammonia, and nourish themselves.

High fertilizer prices helped their sales. By summer 2021, Pivot Bio had taken in an additional $430 million and tripled its revenue. Their product has been used on more than 1 million acres of farmland, and the area was increasing rapidly. In August, the lush Dyersville, Iowa, cornfield surrounding Major League Baseball's Field of Dreams game, nourished by Pivot Bio's product, was seen by millions of viewers. Over the next ten years, the company projects its technology could prevent $200 billion in environmental damage. With investors such as Bill Gates, it released its fourth commercial product, allowing farmers to replace up to forty pounds per acre of fertilizer with environmentally friendly microbes, and it began marketing an organic popcorn.

Many of these start-ups received vital publicity and connections to investors from a new industry clearinghouse based in Oakland, California, called Synbiobeta, that hosted an annual Global Synthetic Biology Conference. This nonprofit consortium of communicators, scientists, and investors was founded by former astrobiologist John Cumbers, who wrote a monthly column in *Forbes* and hosted numerous meetings to call attention to the companies of synthetic biology. It provided a networking platform for those interested in learning more and perhaps investing in the field.

That twin-billed purpose, of doing good while making money, brings us to a final area of interest in synthetic biology's environment focus, something microbes do all the time naturally, cleaning waste.

SLUDGE AS A RESOURCE

Several companies were modifying microbes to clean food, sewage, and agricultural waste and turn it into chemicals for fertilizer and energy. In agriculture, giants like Monsanto, Bayer, and Novozymes were among those designing microbes for that recycling purpose. The Israel-based company Emefcy generated electricity from wastewater by using synthetically engineered microbes.[34] Waste2Watergy was a small company trying to do much the same in Oregon for brewery waste, and, from India,

String Bio was building food resiliency from the methane gas of decomposing agricultural waste like stems, leaves, and roots.

One other area of hope was in cleaning up or degrading plastic. Nature did it, but not fast enough to satisfy demand. A leader in the new industry was Novoloop in California, which used engineered microbes to degrade plastics into industrial, consumer and pharmaceutical product chemicals. The company signed a deal with the town of San Jose to recycle its municipal waste, showing how we may one day change the way plastic is recycled.

As climate change increased temperatures and storm severity, one solution lay in the soil, suggested the synthetic biology company Carbo Culture, based in Finland and which was competing to channel the tons of carbon in plant waste into a compound it can bury in the dirt. The California company NovoNutrients was taking atmospheric and industrial carbon and hydrogen and turning it into fishmeal, the food used in aquaculture, and food proteins for animals and humans..

Notwithstanding such progress, a sign clouded the horizon. Over the past century there has been a frightening increase in outbreaks of spillover diseases moving from animals to humans.[35] Some 60 percent of emerging diseases began with spillover events, the number of which has tripled in the past decade.[36] In 2003, the severe acute respiratory syndrome (SARS-CoV) virus broke out in China, and in 2012 the Middle East respiratory syndrome (MERS-CoV) virus struck Saudi Arabia. Bats were the sources of those two respiratory viruses. We had invaded their remote caves, sometimes to harvest their valuable waste guano as fertilizer.

Motoring my boat slowly beneath the low bridge on the way home, I looked for the bats preparing to set out into the night. Yes to mangroves, I thought, and recycling farm waste and making new kinds of fertilizer replacements, and "probably" to gene editing of invasive mosquitoes if tests continue to show no unforeseen effects. "No" to resurrecting the mammoth. Still, hopes were riding on the potential of lab editing of genetic material to help save Earth. Few realized how important some of that work would prove to be.

8

HEARTH AND HOME

The best technology on planet Earth is nature.
—Eben Bayer, Ecovative

W hen Beyond Meat founder Ethan Brown finished college, his professor father asked him, "What do you want to do with your life?" He had no answer. "What's the biggest problem facing the world?" his father probed further.[1] Climate change, Brown thought. As he looked into the factors changing climate, he found that livestock production was a leading contributor, with its intense use of land and water and warming methane pollution from cattle. "MBAs and engineers would go on and on about fuel cells," Ethan Brown recalled wryly, "and then go out and have a steak."

By 2020, synthetic biology was helping to make sustainable food, drink, and clothing, offering the promise of mediating some of the damage caused by agriculture, responsible for devouring hundreds of thousands of acres and millions of gallons of water, and pouring warming methane into the atmosphere. From burgers to steaks, cranberries to apples, leather to cotton, and the shipping boxes containing them, synthetic biology was seeking to change what we eat and wear. This science brought in idealistic and sometimes conflicted people determined to unravel the tangled threads of cellular processes and, by doing so, to restore Earth.

Of these hearth and home companies, plant-based Beyond Meat was the first success, though it did not use synthetic biology. Since 2008, Brown's Los Angeles–area company had been engineering patties of meatless chopped beef made from plant proteins, apple extracts, and canola oil. Stanford professor Patrick Brown's (no relation) Impossible Burger quickly followed by using yeast to ferment a blood-like soy protein, making its burger more squarely a product of genetic modification than Beyond Meat's plant-based product. While Beyond Meat went public in 2019, the Impossible Burger was being featured in 17,000 restaurants from the United States to Singapore, on *CBS This Morning*, and in newspapers and magazines. By 2020, meatless meat was coming into its own.

Other companies made similar efforts in fishless seafood, cow-free dairy, and in household items like packaging, clothing, and, even, furniture. Boston-based Motif FoodWorks made a variety of different animal-less dairy products and meats. "Try the meatballs!" CEO Jonathan McIntyre enthusiastically urged me at a Ginkgo Bioworks conference cocktail reception. From Ecovative in New York State came packing materials made from mushroom root systems, the same source tapped by clothing manufacturer Bolt Threads and others to replace leather. The seafood company AquaBounty created the first FDA-approved gene-edited fish, a fast-growing Atlantic salmon. In December 2020, a gene-edited pig from Maryland was approved by the FDA: GalSafe pigs created less waste and used less land and water than their conventional cousins.

Companies like Spiber in Tsuruoka, Japan, and Modern Meadow in Brooklyn and New Jersey produced sustainable materials. Berkeley Yeast made beer without water-guzzling hops, instead transferring hops genes into their brewing yeast. The Netherlands-based company DSM sold alternative foods for both animals and humans. California's Finless Foods made seafood without fish, and Perfect Day made milk without cows.

Despite the successes, most products were more expensive and less tasty than the traditional products they sought to replace. Still to be answered was, were they in fact healthier or truly more sustainable. As former academic researchers who started the businesses evolved into new roles as promoters and industrial managers, some critics wondered whether venture capital investment was the best source of funding for a new industry. Small countries with tiny agricultural sectors, such as Israel or Singapore, lacking land for ranching, poured government money into the industries. Calling itself "start-up nation," Israel sought be a leader in the field.

While I sautéed my Morningstar meatless breakfast sausage, the tools of synthetic biology were being applied to sustainable beef, chicken, fish, cheese, fruits and vegetables, clothing, and housing materials, inspiring alternative methods of production and design. The meatless food industry, in particular, had been ignited. "There has been precious little innovation in growing food for a thousand years," Motif's McIntyre liked to say. "That's about to change."[2]

THE MEATY HISTORY OF MEATLESS MEAT

The history of plant-based meat reaches back to 1901. John Harvey Kellogg followed up on corn flakes with a vegetable-based meat called Proteose, which did not catch on in the same way. During World War II, rationing led people to canned substitutes like Choplets made by Seventh Day Adventists. The vegetarian movement introduced in the 1990s the Boca Burger, its chunks of vegetables visible within the gristle.[3] The meat substitute Quorn appeared in the 1990s and is today sold as a fried appetizer, along with the company's meatless chicken, turkey roast, and Salisbury steak.

Then came Beyond Meat. When he was growing up outside of Washington, D.C., Ethan Brown worked weekends on his father's Virginia farm, where he got to know some of the cows he milked by name. There, Brown reflected on the slaughtering of animals and wondered why Americans consume protein the way most of us do, as meat.[4] After earning a BA in history and government from Connecticut College, then a master's in public policy at the University of Maryland in his father's program, he began thinking about global environmental issues. Feeling he needed business preparation to make a difference in the world, he got an MBA at Columbia University and joined a fuel-cell company, promoting hydrogen as a clean gas alternative for cars.

Like the young Steve Jobs, Brown idolized Edwin Land, inventor of the Land camera, whose goal was to create a life-changing product before people realized they needed it. Land promoted products like an artist-evangelist leading society into new ways of seeing. Ethan Brown decided he would do the same for food. In 2009, he revisited the question he had asked himself on his father's farm. His answer was to launch Beyond Meat in a warehouse district in California.

Pea and rice proteins made the stuff of the company's substitute-meat products. Within four years it had produced chicken strips, and by 2017 Beyond Meat scientists created the first draft of its hamburger. Taking the company public in 2019, Brown was more poetic than self-congratulatory. "Let us find out what will beautify the world," he quoted Land, "although people may not know it."

As many readers know, meat agriculture is a wasteful and cruel enterprise, requiring tens of thousands of acres for growing and grazing cattle, terrific amounts of feed corn that dominate American farm planting, hundreds of thousands of gallons of freshwater, not to mention the imprisonment and slaughter of untold numbers of animals. Over the past twenty-five years, forests the size of South America have been razed to create more grazing pasture.[5] Feed cattle contribute an estimated 25 percent of the world's methane gas (a prime object of concern for its global warming effect), emitted through the belches caused by the working of the gut microbes that ferment the grass the cattle eat. Meat production is blamed for a few million tons of waste per year, most of it nitrogen-rich stool and slop stored in foul-smelling pools euphemistically labeled "agricultural runoff." This runoff pollutes rivers and seas and contributes to fish kills in lethal, oxygen-poor red tides. Even a 20 percent replacement of meat consumption by microbial-made proteins could halve the rate of deforestation, according to a 2022 *Nature* study. Our meat excesses damage our health, climate, and natural resources.[6]

The problem is, meat provides sustenance. Humans began consuming raw meat some 3 million years ago and learned to cook it 500,000 years ago. There was no looking back. Grilled, sautéed, smoked, and roasted, meat produced protein energy and delicious calories that enabled human brains to expand and jaws and stomachs to shrink. Meat remains an efficient source of vitamins, and, in malnourished countries, the main concern is to get more protein into people, not less. The meat industry argues that today's yields require one-third fewer cows now than forty years ago, according to the Global Alliance for Climate Smart Agriculture, a pro-business think tank.[7] "We will use the enlarged brains that meat gave us," Brown liked to counter in presentations, "to get us off of it!"

Consumers also questioned the nutritional benefit and agricultural footprint of plant-based burgers. The Beyond Meat burger contained about the same caloric and salt content as a meat burger, and its plant ingredients required thousands of acres of land to produce. Its proteins come from processed rice, peas, and mung beans, not from gene editing, so it is a natural product. To improve on the Beyond Meat agricultural footprint required gene editing and fermentation, which is what Impossible Foods did.

The early leader in genetically modified meat production, Impossible Foods was founded by Stanford biochemistry professor Patrick Brown in 2011 after he took an eighteen-month sabbatical to study the techniques of having microbes ferment proteins. Brown was a science radical who helped found the free online journal *Public Library of Science* in 2001 to circumvent expensive journals like *Cell, Nature, Science,* and specialty publications. His company's 2019 market report had the working title "Fuck the Meat Industry."

The key insight to the Impossible Burger is that meat's appeal comes from blood. Impossible's main ingredient is a soy protein called leghemoglobin, which, in grabbing oxygen from the air, turns itself bright red. The related protein hemoglobin puts the red in the blood of animals, hence in meat. Finding that the DNA sequence making hemoglobin in feed animals is very similar to the DNA sequence of leghemoglobin in soybeans, company scientists worked for years to transfer that soy DNA into yeast. Thousands of failed attempts later, they succeeded in transplanting the soy genes into the yeast strain *Pichia pastoris,* the same strain used by Bolt Threads in making proteins for spider silk fibers.

Their Impossible Burger debuted in July 2016 to fairly good reviews but was high in sodium and somewhat low in protein. A second version, Impossible Burger 2.0, was launched in 2019, lower in sodium and higher in protein but with no cholesterol.

It claimed to use 96 percent less land, 87 percent less water, and generate 89 percent less greenhouse gas than did conventional beef production. It had about the same amount of calories as four ounces of chopped meat. The coconut oil it featured was high in fat, however, and was later replaced with lower-fat sunflower oil, which also sizzled better on the grill.[8]

Questions were raised, however, about the yeast-produced leghemoglobin. Soy contains estrogen-like hormones called isoflavones, which some studies suggested could promote the growth of cancer cells, or impair female fertility, or change men's hormones.[9] But the isoflavone claim was disputed. It was not clear how genes encoding the leghemoglobin protein in soy would cause yeast to produce isoflavones. "Soy does not produce man boobs!" Patrick Brown replied to the critics.

Although priced three times higher than the cheapest ground chuck, the Impossible Burger and Beyond Meat burgers have seen their sales skyrocket, with Impossible Burger making it into restaurants in many countries and Beyond Meat into superstores. Both companies vowed to bring their prices down and the quality up as sales leveled off. Burgers were, however, just the beginning. Other companies turned to steak, ribs, and chops.

Several companies began to study the production of meat and fish cells in the lab to make products using three-dimensional (3-D) printers and real animal cells, not vegetable DNA and proteins, in a process called cell-free or lab-cultured meat or seafood. We turn next to cell-free food.

"A MENU OF ABSTRACTIONS"

In a sleek lab, a huge, whitish scaffold feature is made from plant cellulose. On it grows meat cells, made of real ingredients. Other natural-product scaffolds could be soy protein or vegetable material, and a bioreactor similar to a 3-D printer is churning out huge numbers of cells at high densities. The starting point is stem cells from cows or other animals, these cells cultured in the lab in amino acids. One muscle stem cell can grow up to a trillion meat cells.[10] These grow into a tissue, spliced into strands that also grow rapidly. The resulting meat is paler and less fatty than the original but can be augmented for better look and taste.

Aleph Farms, a leader in the field of cultured meat, was founded in 2017 in Rehovot, Israel. Aleph was creating sustainable steaks with zero waste and could even make food in outer space for long missions as did sustainable seafood company Finless Foods. Aleph's affable Neta Lavon, vice president of research and development, explained that its 3-D printed steak was created by cells arranged in gigantic trays, building muscle and fat tissues. These products are placed on an organic scaffold and coated with canola oil to make them sizzle on the grill. The scaffold material is critical because cells are living machines, and the organic ladders affect how well the cells will function as factories. Cell-free is "lifting the hood of the car, pulling out the engine, and repurposing it for other things," said Northwestern University's Michael Jewett, who became a director in 2022 of a joint U.S. Army institute for cell-free production of military materials like camouflage and tents.[11] By November 2022, the US FDA declared that cell free, or slaughter free, chicken made by the California company Good Meat, a division of UPSIDE Foods, was safe to eat.[12]

Other companies were racing to expand cell-free manufacturing of other foods. California-basedBlueNalu was producing cell-based seafood. In New York, Atlast Foods came out with plant-based bacon, while the New York State–based company Ecovative was making bacon out of the root systems of mushrooms. The processes were becoming so sophisticated some lab-grown companies rejected the designation of alternative. "We make meat, not meat alternatives," said the brash Joshua March-Henderson, cofounder and CEO of cell-free company Artemys Foods in San Francisco.

Some founders were devotees of the hyper-local food movement, in which highly trained chefs established ties with local farmers to bring diners to directly experience their food. "We are not in touch with what we eat," explained Patricia Bubner, an Austrian-born scientist who cofounded San Francisco–based Orbillion Bio, which was replacing wagyu beef with cell-based meat. "Growing up, I knew the farmer who raised the cows for our milk and chickens for our eggs, or the plants for our vegetables. I came to Berkeley for my postdoc and became entranced by the local food movement," she said. To EatJust cofounder Josh Tetrick, fast food from frozen-and-shipped meat is "a menu of abstractions."

By then, food synthetic biology companies explored new techniques in locations all around the world. UPSIDE Foods in Berkeley, Lightlife in Toronto and Massachusetts, Field Roast in Seattle, Utrent in San Diego, and Sweet Earth in northern California were finding success with substitutes such as mushroom root systems as an ingredient for MyBacon, later purchased by Ecovative. Several other companies were bought by industry behemoths, such as Nestle buying Sweet Earth and Future Meat Technologies.[13] Tyson invested in Beyond Meat. Archer Daniels Midland partnered with Brazil-based Marfrig to sell meat-free Revolution burgers. Cargill backed the precursor to Upside Foods. Maple Leaf Foods acquired Lightlife and sold its products in Canada.

Sustainable food spread to big chain stores. Walmart sold Beyond Meat hot Italian sausage. Trader Joe's offered affordable Soy Chorizo and Chicken-less Strips. Target featured Ripple, the pea protein–based milk, and its own line of plant-based meats called Good & Gather. Meatless meat was projected to grow into a $25 billion market by 2030 as more giant manufacturers skipped the start-ups to start their own divisions.[14] Chipotle advertises its plant-based chorizo, but it was only available for a limited time.

Sustainability came from the farm as well. In Canada, the Enviropig was being genetically modified for fast growth, disease resistance, and decreased phosphorus in its manure. The Enviropig gene complex came from a mouse and from *E. coli*.[15]

Close behind meat and seafood came candies. One key snack ingredient was collagen, made by grinding up the skin, bones, and tendons of animals into gelatin, to go into favorite candies like Gummi Bears or Starburst. "That process hasn't changed since the seventeenth century!" Stephanie Michelsen, the company Jellatech's quipping CEO, exclaimed at a conference.[16] Jellatech moved the genetic pathways found in natural collagen into laboratory cells to make gelatin building-blocks. The company envisioned soaring sales ahead for sustainable gummi vitamins and cosmetics.[17]

Next up is sustainable clothing, seeking to reform an industry notorious for its global waste, animal cruelty, and poor working conditions.

FROM HEARTH TO HOME

In 2018, the sustainable clothing company Bolt Threads began a historic collaboration with British designer Stella McCartney, and every one of the ties, hats, and other clothing items they made sold out, even at their high prices. McCartney wanted more. She was committed to ethical production, and a main concern of consumers and designers was the sacrifice of animals for leather. Bolt Threads' chief technology officer, David Breslauer, told me, "Clients kept asking us, 'Can you make leather?'"

The solution was to use as the raw material mycelia, the feathery mushroom root systems that can be grown into material for fabric in a short time. A century ago, East Europeans had known that fine-spun mushroom root systems could be crafted into beautiful, leatherlike fabrics. The company Mycoworks pioneered the use of mycelia to make a leather substitute it called Reishi. Now Bolt Threads' researchers sought to take that knowledge further.

It required hundreds of different formulations before Bolt Threads realized the leather substitute they called Mylo, Breslauer told me. The leather substitute originated in the threadlike mycelium roots of mushrooms found beneath plants and trees (mushrooms are the fruit of mycelia in the same way grapes are the fruit of vines). Mylo began as cells growing into a foamy layer like squished marshmallows in eight to ten days, feeding off of sawdust as a nutrient. The layers are refined and dyed and the waste composted. This process was much faster and more sustainable than the eighteen months to two years for a cow to develop. In 2018, Stella McCartney produced her bestselling Falabella black leather purses out of Mylo, which was produced in factory facilities in Arnhem, the Netherlands; Portland, Oregon; and Emeryville, California.

The company in 2022 was creating a new production site in the Netherlands because "we're going to have to grow that site at a ridiculous pace to keep up with demand," CEO Dan Widmaier told me. Demand for Mylo increased through agreements with Adidas, to make the formerly leather uppers of its Stan Smith tennis shoes, Lululemon, for yoga pants, and luxury clothing giant Kering. In 2022, Mercedes Benz came out with Mylo interiors for its new electric Vision EQXX sedan. When I asked Dan Widmaier whether, in graduate school, he anticipated hanging out with Paris designers, he said: "No. No, no! The fashion part, no way!"

As he reminisced, Widmaier considered the potential global impact of developing a company, as opposed to remaining in a university research lab. "We can quibble about whether synthetic biology is a rebranding of the same-old genetic engineering we'd been doing for twenty years, but for the ones who really embraced the principles of engineering biology for a wildly different, techno-forward future . . . it's really hard to solve big societal problems with academia alone," he said.[18]

For her part, Stella McCartney featured a Mylo body suit in Paris. "This," she told the *New York Times*, "is going to change everything."[19]

Mycelia also provided the basis for sustainable packing materials. An estimated 20 percent of the volume of U.S. landfills is taken up by expanded Styrofoam. Since one cubic foot of Styrofoam requires the expenditure of 1½ liters of petroleum, tremendous amounts of energy are wasted in making it, and more in breaking it down. Moreover, Styrofoam itself can remain in the environment for up to five centuries, and less than 1 percent gets recycled. As online sales of home-delivered goods skyrocketed during the COVID-19 pandemic, so did the volume of packaging waste. The items bought online were shipped in boxes packed with Styrofoam filler.

Ecovative cofounder Eben Bayer's familiarity with the versatility and strength of mycelium began with splitting rotted logs at his family's Vermont farm, an arduous task made harder by mycelium's strength binding the wood. Tall in his hiking boots and a swooping, skater-boy haircut, Bayer, a Rensselaer

Polytechnic Institute alumnus, recalled, "Pigs and chickens were really cool. . . . (But fungus) is compostable, circular, evolved over millions of years."[20] At Ecovative, researchers produced mycelium-based substitutes with similar properties to the packing material, but which could decompose within only forty-five days. The company claims the sustainable packaging is better in quality than the original.

The process of creating products with mycelia starts with crop waste—wood chips, corn husks, hemp, plant stocks—anything discarded in growing crops. The waste is placed in designed molds of recyclable plastic sheets and inoculated with mycelia, which performs like a glue. Water is added, and the mixture is packed into a mold, where the mycelium cells feed on the waste and grow into something resembling the same huge, white marshmallows as in Bolt Threads' process. In both cases, the mycelium itself did not require synthetic biology. Ecovative's main product debuted as packaging inside shipping boxes holding tech giant Dell's servers. Bayer explained, "We want people to adopt this technology because it's cheaper and it performs better than plastics. Our sneaky run-around is that it's actually better for the planet."[21] In a more recent innovation, Ecovative built a tiny house with walls of mycelia. By 2022, the company spun off one subsidiary called Forager that manufactured vegan materials for fashion and automobiles, and another called MyForest Foods that made MyBacon, which I sampled at the 2022 Built with Biology conference. The bacon was not bad.

Another source of global waste is discarded clothing. The Swedish company Renewcell engineered microbes to recycle discarded clothing into a clean, easy-to-sew material that it called Circulose. Its first proof-of-concept in 2020 was a dress made for the Swedish clothing company H&M that was 50 percent Circulose. At a conference in New York's Hilton Hotel, I held a gray, silky T-shirt from another creator of substitute textiles, the Japanese company Spiber. The high-end shirt contained a fraction of real silk but was mostly made with yeast-fermented proteins. It felt as cool and sleek as water in my hands. With ingredients for blue jean indigo produced by modified bacteria, a new age of sustainable materials was struggling to take hold, bringing me back to the sea.

NOT YOUR FATHER'S SURF AND TURF

As challenging as it was to alter agriculture, it was almost harder to confront the dangers of ocean warming and overfishing. Fueled by nutritionists' warnings about the hazards of eating red meat, fish consumption rose 400 percent from 1970 to 2019 in the United States, and overfishing became a global problem. Fishing restrictions were notoriously difficult to enforce, and consumers' global demand for shark fin soup put those predators on the endangered species lists, along with numerous other fish.

One of the first fish-alternative companies was Finless Foods, established in California in 2017. Its goal of making an alternative tuna steak began with a simpler fish-free sashimi. "That's just muscle-fat, muscle-fat," said cofounder Michael Selden, as he recounted how the company[22] achieved the desired taste. The cells were placed in a bioreactor and fed salts, sugar, and proteins, replicating natural processes that occur inside of an animal. The replication is followed by placing the cells on a scaffold to assemble in a predetermined 3-D structure.

Finless Foods was also reaching for the stars. Partnering with a Russian mission and a company called 3D Bioprinting Solutions, in 2019 it sent cells and equipment into outer space to grow cell-free seafood aboard the International Space Station.

They argued the experiment would save money and precious payload weight, while providing better meals than the standard prepackaged astronaut food.

Then there was seaweed. All green plants on land can trace their origins to the sea plant. As Japanese, Koreans, and Chinese consumers had long known, "seaweed has the potential to address world hunger, climate change, and biodiversity," the tall, United Nations global compact senior advisor Vincent Doumeizel said in conference speeches. Doumeizel foresaw a coming "seaweed revolution" including carbon-absorbing seaweed farms whose produce is cured by solar-powered passive beach dryers. Of the three most commonly known types of seaweed—green, red, and brown—green seaweed is the origin of land plants and closer genetically to an oak than to the other two kinds of seaweed. Today we cultivate only a few seaweed species out of some 35,000. "We have such ignorance," Doumeizel exclaimed, "we are like the farmers starting agriculture 12,000 years ago!"

The advances in engineering cells to make a variety of hearth and home products were accelerating, including algae to make edible proteins, opening the way for visionaries to mimic nature's incredible creativity in art.

GARDENERS OF EDEN

An enduring interest in biodesign, the use of natural forms in man-made works of art, appears in prehistoric French cave paintings and the masks and intricate art works of pre-Columbian American Indians. By the 1990s, modern biology–inspired art exhibitions were springing up in London, Paris, and New York, making use of natural materials to mimic plant and shell forms. A nonprofit, international BioDesign Challenge competition encouraged high school and college students to devise engineering projects for awards. The 2022 finalists included a Japanese student team's "post-anthropocentric vending machine," a Universidad de los Andes team's moss water filter, and a California College of the Arts student's mosquito net made from lemon rinds.[23]

Another biodesign pioneer was English artist and author Alexandra Daisy Ginsberg, whose English countryside installation, Pollinator Pathmaker, featured colorful flowers to provide havens for native bees and butterflies under threat. Other features included an interactive website with which viewers create their own visions of what Ginsberg calls The Eden Project.

Within the realm of house products, the potential of life's variety seemed boundless. Bolt Threads considered harnessing the adhesive powers of mussels and barnacles for glues. Other companies studied the hard structure of shells for construction materials. At Rice University, Caroline Ajo-Franklin and her lab were mimicking seashell structures and colors. "Their iridescence comes from its organic layers of calcium," noted Ajo-Franklin, whose research in biosensors had applications in carbon capture, bioenergy, and assembly of tiny structures. DARPA's Engineered Living Materials fund provided an initial grant. In one project, her lab mimicked the intricate chain armor of a slime-forming lake bacterium, *Caulobacter crescentus*.

Some products tapped a microbial survival mechanism dating to ancient life: biofilms. These sticky protective coatings of bacteria on surfaces like metal or rock are created by secretions to make a tenacious toehold for life. Biofilms can help treat wastewater and remove pollutants and make antibiotics and vitamins.[24] Biology may also form its own energy sources, like an Israeli desert wasp that carries a photovoltaic cell on its back. A threatened octopus or squid can change its color and texture by using chromatophores, color-changing cells, in its skin. Damaged papaya fruit produces latex and heals itself. In North Carolina, the Biomason company created concrete by making use of bacteria, crushed rock or sand, nutrients and calcium and carbon sources. The process resembled the ways corals and sea creatures created reefs.

Utilizing nature remained challenging, however. Bioplastics, for instance, required use of more land than in the traditional process of making plastics from petroleum. But interest in the power of synthetic biology came with sights set on reforming humans' relationship with nature from one of domination to one of cooperation. "We will eventually grow materials in a seamless interface between humans and nature," Ajo-Franklin predicted.

A final vantage point to see that interface was in easy-to-eat fruits and vegetables.

"KNOW WHAT YOU'RE DOING"

On a drive to a Massachusetts farm, researchers Tom Adams and Haven Baker were musing how fruits and vegetables needed to be easier to snack on. Thoughts of pitless cherries, seedless blackberries, or mangoes and pineapples the shape and size of seedless grapes led them to found Pairwise outside Raleigh-Durham, North Carolina. They began with a goal of improving the bitter taste of mustard greens. "If we change the taste," said Pairwise science director Haven Baker, "we have a leafy lettuce more nutritious and much cheaper than kale."

Under CEO and Monsanto alumnus Adams, Pairwise partnered with German corporation Bayer to improve the hardiness of licensed row crops—corn, soybean, wheat, cotton, and canola varieties. They devised a method to increase the pest resistance of cranberries by tapping the best qualities of the fruit's hundreds of variants, of which only two were farmed in the United States. Adams was leading a 125-person company, whose science board included CRISPR pioneer Feng Zhang from Harvard, and which drew early investments of $125 million from Bayer and another $120 million from investors in Singapore and New York.[25]

Augmenting fruits and vegetables was capturing the imaginations of others too. Norwich-based Tropic Biosciences in England was editing banana genes to help bananas resist a widespread fungal disease. Also in the England, John Innes Centre researchers were working on nutritious "super broccoli," the first approved European GMO, with a plan of having it on the market by 2023.[26] Agriculture giant Conagen was partnering with small flavor-and-fragrance ingredients company Blue California to make vanilla, citrus, and floral flavorings through fermentation. It already had a variety of products like musk perfume, natural preservatives, synthetic vanillin, and industrial chemicals for sale. The scope of research that began with Beyond Meat, dubbed plant-based 2.0, advanced to a new era dubbed plant-based 3.0.

The state of the bioindustry in 2022 stood at several hundred synthetic biology companies, compared to fewer than one hundred a few years earlier. Investments were increasing. Early investors included former Google CEO Eric Schmidt, whose capital firm had invested in Bolt Threads, Zymergen, and Ukko, the latter synthetically engineering meals to help people with food allergies. PayPal founder Peter Thiel was also invested in Bolt Threads and in a remote controlled biology company called Emerald Cloud. Netscape co-creator Marc Andreessen invested in Benchling, a company in which biologists could engineer DNA with software before turning to experiments in the lab.[27]

Two of the most successful synthetic biology product categories were cosmetics and fragrances. Through yeast fermentation, companies like Ginkgo Bioworks provided the ingredients for cosmetics giants such as Robertet, but clients remained secretive about their products. In 2021, Ginkgo partnered to do the same with flavor manufacturer Givaudan.[28] Bolt Threads created a skin-care product called B-silk, sold in a Los Angeles–based hair-care line, and hired Ginkgo to help optimize its Mylo manufacturing. Conagen made cosmetics and food products.

Generally, though, the market shares remained small. None of the giant synthetic biology companies were turning a profit. Even some of the health and green claims of synthetic biology manufacturers were challenged. In clothing, for the moment, it seemed like the most successful garments were only partially created by synthetic biology. Zara's attractive, little black $69 carbon exhaust dress, made in partnership with LanzaTech, was for sale online for Christmas 2021. As I marveled at the airy lightness of a gray, partially synthetic Spiber T-shirt, I tried to order the company's synthetic spider silk, cream-colored Moon Parka ski jacket. However, only fifty were produced. Adidas's Stan Smith Mylo-upper tennis shoes were still in production. While I waited to order my sustainable clothes, the field of synthetic biology was applied to mining and the military, both in heightened states of alert.

9

FANTASTIC VOYAGES
Mining and the Military

It's one of those rare moments . . . of an exponential curve in both discovery and tool making.

—George Church

From the high plateau of Chile's Atacama Desert, gigantic trucks are moving three-story-high mounds of copper and rock.[1] Bacteria are placed on top of the waste rock and left to break it down. Welcome to biomining, the process of using microbes to extract metals from rocks, which came to prominence in copper mining as a synthetic biology solution to environmental issues and played-out mines. Copper smelting created arsenic and sulfuric acid. Designed microbes ate both. Some forty-four companies, in countries ranging from Kazakhstan to India to Ghana, were using biomining practices, not only for copper, but also for other valuable metals and minerals. The world's biggest bioreactors could even clean up garbage dumps.

For centuries, pickaxes and then giant machines had been used to extract Earth's minerals, but the mining industry was notoriously wasteful and damaging to the land. While mining accounted for some 4 to 7 percent of humanity's carbon dioxide emissions, it produced tons of toxic runoff and waste, devastating the countryside with its gigantic excavations.[2] Yet, precious metals are more valuable than ever, central to technologies that address the climate crisis. Synthetic biologists believed the future of sustainable mining naturally existed in the ground.

Deploying microbes to pull out gold, copper, uranium, and nickel from waste heaps and to clean them at the same time, biomining could be a profitable business. Traditional mining methods recover 60 percent of a mine's valuable ore; biomining could extract 90 percent. For this reason, biomining has been called the "mining of the future."[3]

The Department of Defense was very interested. Similar microbe use and engineering could help the U.S. military recycle tons of waste metals such as brass, steel, and iron into biofuels and feedstock. Bacterial bomb sensors and performance enhancers for the modern soldier were the goals of several secret programs around the world. China and the United Kingdom were making synthetic biology a cornerstone of their national military improvements. Vladimir Putin wrote about new genetic weapons while quietly expanding Russian programs to synthetically engineer smarter, stronger toxins.

Bioterror emerged as a bigger threat than ever, as arguments raged over the origin of COVID-19. In previous years, Iran, North Korea, and rogue groups had been considered the prime bad actors, but China, Russia and the United States had also researched the use of pathogens as bioweapons. The feared weapons were not only warheads, but bacterial infiltrations into the water or food supply or the leaking of enhanced viruses. In the past, a group of researchers was criticized for synthesizing a smallpox variant, then publishing the technique.[4] Now, the U.S. Intelligence Advanced Research Projects Activity (IARPA) funds labs to create algorithms to spot newly engineered killer pathogens before they attack.[5] But nothing was more emblematic of biology's disruptive power than the COVID-19 pandemic.

In the United States, DARPA had long given critical help to synthetic biology research at early stages of development, including ideas on biosecurity. "We want to avoid the mistakes of the cyberworld," suggested former defense assistant secretary Andrew Weber, referring to the daily attacks of virus spyware since the advents of the personal computer and the internet. "Failing to fund security at the outset, it has been playing catchup ever since," Weber said.[6]

A synthetic biology industry seemed finally to be gaining ground. Toward that end, DARPA funded innovative developments for use in heavy industry and the military. "It's one of those rare moments, unprecedented or partially precedented, of an exponential curve in discovery and tool making," Harvard's George Church told me. "It has the feel of the Edison electrical revolution, but this is going faster."[7]

From China, a respiratory virus spread around the world in 2020 and forced governments to respond. As public health systems raced against time, researchers struggled to understand how their discoveries might be misused to cause future pandemics. To understand the militarization of biology, one could start with the past.

A RECORD OF BIOWARFARE

In ancient Rome, armies doused articles of clothing with the smallpox virus, and then launched them by catapult against their enemies.[8] During the French and Indian War, British general Jeffery Amherst authorized the provision of smallpox-infected blankets to the Indians of Pennsylvania.[9] The Germans during

World Wars I and II created biofactories for producing poison gas, and the Japanese poisoned the wells of their enemies during World War II by contaminating them with carcasses of diseased cattle.

There was nothing unique about such weapons. Both the United States and the Soviet Union funded huge Cold War programs to engineer pathogens to unleash into cities. Occasionally, deadly accidents occurred. A research institute in Sverdlovsk during the Soviet era sprang an infamous anthrax leak that killed nearly sixty-eight civilians in 1979, by government admission, with estimates of the real death toll ranging up to 1,000 people. If mistakes were made, sometimes they were admitted—most notably by the new Russian Federation government in the case of the Sverdlovsk anthrax leak—but more often they were not. Lab leaks were often shrouded in secrecy and thus hard to document and analyze.

At least one outbreak of a virus was acknowledged as an accidental lab leak, the emergence of influenza A virus in China in May 1977.[10] Genetic sequencing suggested the outbreak was the result of a lab escape of a virus that itself had first been discovered during the period 1948–1950.[11] It was a seasonal virus, affecting children, and fatalities were relatively few, but there is evidence it spread widely in Russia.[12] Shortly afterward came the Soviet release of anthrax in the city now called Yekaterinburg. Some ninety-four people were infected in May 1979 after a worker removed a faulty filter at the end of a Friday and no one on the next shift replaced it.[13] That story, and the story of the two-decades-long Soviet program to enhance smallpox and anthrax as killers, were recounted by the physician and former program director Ken Alibek in his 1999 book *Biohazard*.[14]

A 1995 outbreak of Venezuelan equine encephalitis in Venezuela and Colombia, a virus transmitted from horses, was also likely to be a lab leak from a research facility trying to develop more powerful vaccines, according to a report in the American Society of Microbiology's *Journal of Virology*.[15] Six leaks were documented from the labs studying the 2003 severe acute respiratory syndrome (SARS) coronavirus outbreak, one in Taiwan, one in Singapore, and four from a single lab in Beijing.[16] Altogether they appeared to be responsible for one death, the physician mother of one lab worker who had traveled home by train. In 2007, a British outbreak of foot-and-mouth disease among farm animals was traced to a faulty drainage system at a Pirbright, England, biosafety level 4 lab facility, costing some £200 million in culled livestock.[17] In November 2019, right before the SARS-CoV-2 outbreak, a vaccine factory leak in northwest China resulted in some 10,000 people developing brucellosis (a bacterial disease common in cattle) because of the use of expired disinfectants. The factory was shut down.[18]

Until then, all the previous lab leaks had been of known pathogens, limited in range and low in death toll. No one, to popular knowledge, had leaked a lab-enhanced virus into a city or region. Still, the potential was obvious enough that on the eve of the COVID-19 pandemic, public health officers were proposing an international effort to prepare for a pandemic threat.

A more positive use of synthetic biology could be found in a necessary industry. This is the new business called mining without mines.

MINING WITHOUT MINES

The vast Atacama Desert stretched red as a Mars surface under a faint, cold sun at dawn. Gigantic earth-moving trucks cart away tons of rubble from an enormous mining pit and dump it into slag heaps. But what if microbes could extract valuable ore from those heaps? That is what the Chilean mining company BioSigma S.A. and others are seeking to do. The use of synthetic biology in mining may make a dirty, wasteful process cleaner and more efficient. It leaches valuable ore and reduced carbon emissions from vast waste pits. It does so by using engineered photosynthetic cyanobacteria to remove metals from wastewater and slag containing sulfate and nitrate compounds. The potential is to treat hundreds of millions of gallons of wastewater a year and to pull ore sustainably from abandoned mines.[19]

Biomining most commonly focuses on copper, uranium, nickel, and iron. Other metals, such as gold, can become more accessible through traditional mining techniques if the rocks are dissolved in a process called bioleaching. The same process occurs in nature over hundreds of years; bioleaching simply speeds it up. Today, some 15 percent of copper and 5 percent of gold on Earth comes from biomining, and that percentage is growing. The bacteria are natural, not the products of synthetic biology, but rather chosen for their strange love of toxic environments that made them candidates for life on Mars.[20] Normally reserved for metals that can be recovered as microbes break down waste, biomining is most economical where conventional mining has been completed and a mine played out.[21]

Such was the case in Chile. In that country's copper mines, bioleaching comprises some 10 percent of mining today, and the percentage is growing. BioSigma in Santiago was a joint venture by the Chilean mining company Codelco and Japan's Nippon Metals and Mining. It was almost a closed-loop process, meaning no extra resources are used, just the natural microbes. Pilar Parada Valdecantos, BioSigma's biotechnology manager, pointed out: "These bacteria need very little air and mainly oxygen and CO_2, and use the mineral itself as their source of energy."[22]

Chile provides nearly 30 percent of the world's copper supplies and accounts for more than half of the country's exports, rendering efficient sourcing crucial for both the nation's economy and global stockpiles. Chile's biomining could limit CO_2 emissions, toxic chemical buildup, and dangerous working conditions for miners. As the world's largest copper exporter, its embrace of biomining could have a tremendous impact on making a circular economy in one of the world's most environmentally damaging industries. It worked in a pretty straightforward manner.

One of the three most common biomining methods is heap leaching, in which waste ore is piled into heaps where microbes are added and left to work. Dump leaching, by contrast, consists of low-value ore put in a sealed pit and bioleached. In a newer process called agitation leaching, crushed rocks are placed in large vats and shaken to distribute microbes evenly. The time span required for these biomining processes, ranging from months to weeks, is their limiting factor, making them slower than mechanical extraction.

Still small in proportion to the overall industry, microbial bioleaching was slated to increase in use as mines become depleted and more subject to environmental regulations.[23] In Arizona, several mine leaching pads in the desert subject slag to microbial erosion. Worldwide, some leading companies studying or implementing biomining include Rio Tinto, NQ Minerals, and Apex Minerals. The main biomining countries include South Africa, Brazil, Chile, and Australia, followed to a lesser extent by the United States, where the Houston-based company Cemvita is an innovator. In Germany, the Biotechnology Research and Information Network (BRAIN) maintains a data base of useful biomining microbes and manufactures a modular van-sized BioXtractor that can help companies retrieve precious metals from waste streams.[24] Canada is also active in biomining research and development.

In addition to copper, biomining occurs for other precious minerals. To reveal treasure such as gold, silver, copper, zinc, and uranium within waste heaps and sediments, sulfuric acids are introduced to promote the solubility of metal ions. These are absorbed or accumulated by cultures of bacteria or their microbial cousins called archaea. During this treatment, the internal temperatures of the ore can reach up to 122 degrees Celsius. More dissolving acid can be added manually or produced by the microbes themselves. The deposits are washed to recover the metal-laden microbes for processing.

Another biological option is to craft green plants to search for petroleum and gold. It turns out that plant microbiomes change when they detect these substances in the environment.[25] A eucalyptus tree that detects buried gold will produce gold flecks on its leaves, glittering thoughts "in a green shade," to quote poet Andrew Marvell.

Finally, bioleaching can clean mine waste, thus providing a double benefit.[26] Bacterially derived polymers can be added to promote growth of the remediators, and the microbes can be engineered for greater efficiency. Most of the microbes are acid-loving bacteria or archaea with high heat-tolerance and the ability to oxidize iron- and sulfur-containing compounds. The most commonly used are *Acidithiobacillus*, either solely or in mixed cultures including bacteria such as *Leptospirillum* and *Sulfobacillus*.

The ability to engineer microbes to clean waste also caught the imagination of many branches of the U.S. military, where the grants agencies DARPA and ARPA-Energy had pioneered some of the most significant advances of the past forty years. It is time to understand a little more about these little-known federal agencies and their roles in the growth of synthetic biology as a science.

VIRAL PROTECTION

In the 1990s, DARPA began investing in synthetic biology through grants to academic, corporate, and governmental partners, not only to advance military proficiency, but also to craft developments in data storage, medicine, and alternative energy sources. From its six regional offices employing 220 government employees, 100 program managers supervised some 250 programs. These sponsored research partnerships typically ran three to five years and involved projects in science and systems capabilities directed toward a goal of achieving tangible results.

DARPA's Biology Technologies Office had awarded seed grants to Ginkgo Bioworks, Zymergen, and Twist Bioscience, while its Engineered Living Materials Program focused on programming DNA and RNA. One such effort provided nearly $25 million in 2013 to support Moderna's early search for RNA vaccines. Moderna received additional federal funding from the Biomedical Advanced Research and Development Authority (BARDA), a division of the Office of the Assistant Secretary for Preparedness and Response (ASPR) within the Department of Health and Human Services (HHS).

To advance synthetic biology solutions for security issues, the Department of Defense created its Biological Technologies Office, which in turn initiated programs such as Battlefield Medicine and Safe Genes, supporting research into restorative therapies for injured soldiers, including neural implants for veterans injured by high-concussion bombs. DARPA funded other companies and universities such as MRIGlobal, the J. Craig Venter Institute, MIT, Stanford, California Institute of Technology, the Foundation for Applied Molecular Evolution, and still more. The ultimate goals were large-scale production of sophisticated tools to make bioengineering more efficient and accurate.

By contrast, ARPA-Energy (ARPA-E), a Department of Energy agency, funded biofuel research such as Keasling's continuing efforts at Berkeley. In addition to these programs, aerobic microbes could be used by the military to make ferric iron or to degrade metals like brass, steel, and iron, while anaerobic microbes could recycle waste into fuels and feedstock, helping to handle the massive waste in a military installation or campaign. They could also help protect community infrastructure against attack by sulfate-reducing bacteria, which were a billion-dollar oil pipeline problem.

With the next generation of synthetic biology applications slated to contribute up to $4 trillion to the American economy, the bipartisan-passed Fiscal Year 2022 National Defense Authorization Act (NDAA) capitalized on the early efforts. It established a National Security Commission on Emerging Biotechnology to implement policies to advance the development of synthetic biology and also protect the country against biological threats. A federal biotechnology plan was a critical policy recommendation of the White House Executive Order on Advancing Biotechnology in 2022. As Heidi Shyu, Under Secretary of Defense for Research and Engineering put it in September, 2022, the Defense Department "recognizes biotechnology as a Critical Defense Area that will change the way DoD develops new capabilities, conducts missions, and adapts to major global challenges."[27]

One big challenge of the future is to stop pandemics before they happen. Toward that end, IARPA commissioned five labs to create algorithms capable of spotting killer viral genetic sequences. DNA makers like Twist Bioscience invested millions for screening malevolent actors who might order pathogenic sequences. Several other pandemic defense efforts were picking up momentum, including those run by the Departments of

Health and Human Services, National Institutes of Health and National Institute of Allergy and Infectious Disease. Then the U.S. Air Force became interested.

"THE FUTURE IS BRIGHT"

From a packed conference room at the New York Hilton I listen raptly, typing in my laptop. "I'm a chemist making hybrid synthetic cells," brown-haired, thoughtful Nancy Kelley-Loughnane is saying. At the Air Force Research Lab (AFRL) in Dayton, Ohio, she and others had been urging the air force to fund more research into synthetic biology. "The future of material synthesis for military environments is bright," commented Kelley-Loughnane, describing her study of more efficient methods of bioproduction of materials.

Synthetic biology might help reduce the size and weight of military equipment and improve biosensors or infrastructure in hostile environments. Kelly-Loughnane's research concerned, among other things, Nvjp-1, a sandworm jaw protein that makes up the animal's strong jaw and increases hardness. "We think that can be used for treatment of soldiers who've lost limbs to IEDs," she said. Coordinating the activities of more than 100 scientists and engineers from the Army, Navy and Air Force, she worked on teams that were looking at high-energy density propellants, which could be made at lower cost and higher energy density than petroleum-based fuels, as well as a number of other applications.[28] The Air Force Research Lab had, among other projects, engineered a gut microbe to produce a beneficial metabolite when it sensed stress in a pilot's body.[29]

DARPA had even funded partners in artificial intelligence. As mentioned in chapter 5, DNA is a more stable digital storage medium than silicon, lasting hundreds of thousands of years as opposed to a couple of decades. One could translate digital data from binary o's and 1's into the A's, C's, T's, and G's of our genes. In addition, bioengineering itself creates huge data sets that need standardization, with machine learning and automated lab protocols improving its speed and calming its noise. Yvonne Linney, former CEO of the company Strateos, pointed out that after labs closed during the COVID-19 pandemic, translation of lab protocols into code enabled scientists to run their experiments remotely.[30]

Along with the military's interest in synthetic biology as a data storer is the prospect of bioremediation. How could synthetic biology engineer microbes to recycle waste? To answer that, one had to spend time in India, Australia, and in New York State.

CLEANING BROWNFIELDS WITH BIOLOGY

Stony Brook, New York, is home to Allied Microbiota founded by Ray Sambrotto, the company's president and chief science officer as well as a Columbia University professor. The company developed a toxin-eating microbe to clean up brownfield contaminants like polychlorinated biphenyls (PCBs), dioxins, and polyaromatic hydrocarbons (PAHs) in the soils and sediments of industrial sites. Founded in 2017, Allied Microbiota employed natural microbes but put them on a special patented diet to speed up and lower the cost of cleanup processes, replacing the trucking and re-dumping of contaminated dirt hundreds or thousands of miles away.

Across the world in Southeast Asia, India was a leader in the use of bioremediation to remediate trash heaps. At Kumbakonam in southern India, where the numerous universities earned the town the nickname "the Cambridge of India," the company Zigma Global Environ Solutions has been treating a seven-and-a-half-acre waste heap since 2015, using microbes to break down the trash. Using similar technologies, the rest of India was converting some 12 million tons of waste into soil in 2016.[31]

Australia was focusing on reclamation. Australia's national science agency, the Commonwealth Scientific and Industrial Research Organization (CSIRO), partly based in Perth, funded biotech for improving precious water quality. As CSIRO scientist Anna Kaksonen explained, "Microbes can remediate sulfate, nitrate, or selenite. They can clean organic impurities or make wastewater more acidic or alkaline, as needed."

Bioremediation makes use of plants and bacteria or their cousins archaea to mimic natural systems for cleaning wastewater. Researchers begin the process by analyzing DNA to identify the microbes already contained in toxic industrial or mine water. They then search databases of commercial microbe cultures to conduct biomining experiments themselves, augmenting nature to make microbes that thrive in harsh environments. Amazingly, those searches have led them to identify microbes that create biodiesel chemicals in symbiosis with microbes in wastewater, said Kaksonen. CSIRO-funded researchers were so successful that the company Evolution Mining hired them to treat the sulfate- and metal-contaminated wastewater produced by its gigantic Mt. Rawdon gold mines near Queensland.

The company wanted to create a new wetlands system from sawdust, plant material, two plant-based chemicals, and synthetically designed microbes to reduce sulfate and remove nitrate from waste and extract more ore from it. "Because metals can be more easily recovered from sulfides, and bioprocesses can use organic waste streams, these techniques also reduce operating costs," Kaksonen noted.

Synthetic biology to clean mine-contaminated soil also involves engineering green plants. Phytoremediation, in which metal-accumulating plant species are grown atop the contaminated soil, allows larger areas to be decontaminated more cheaply than by man-made technologies of remediation. Some tree species include poplar and willow for cadmium removal, *Brassica* for chromium, copper, lead, and nickel removal, and *Pteris* for arsenic. Afterward, the plant or its leaves and branches are harvested and composted, removing the contamination.

Scott Banta is a Columbia University professor of chemical engineering interested in bioremediation. His team explores, among other projects, the potential of bacteria in mining remediation, using engineered versions of the bacterium *Acidithiobacillus ferrooxidans* to oxidize iron and sulfur. The researchers' engineered microbes can be grown on iron aerobically or anaerobically, with oxygen or without. The microbes do what they do in the wild, form biofilms that break down metal. "Mining companies feel they are in the forefront of solutions to environmental problems," Banta liked to say in conference presentations.[32] To take one example, Banta notes, copper smelting produced arsenic by the ton. Biological solutions for removing arsenic might include blasting the rocks, treating rubble with sulfuric acid, or applying bacteria to eat the waste and leach out the metal, and the cleaned waste could end up stacked hundreds of feet in the air. One such heap in Chile's Atacama Desert was dubbed "the world's largest bioreactor," Banta recollected.

What would it take to make synthetic biology alter the economy of the 2030s? "Synthetic biology is going to remake the world," concluded one U.S. Air Force report.[33] The COVID-19 response showed how that could happen.

"TRANSFORMATIONAL MOMENT"

In Wuhan, China, in December 2019, people were getting sick with a strange form of pneumonia. When they sneezed, they spread droplets containing a virus with spiky molecules that made protein prongs resembling those on a king's crown, similar to those seen in virus outbreaks in the Middle East in 2009 and earlier, in 2003, in China. In January 2020, a Chinese scientist, at risk to himself, leaked the virus genome on social media, giving researchers a blueprint to study. Researchers at companies such as Moderna, BioNTech, CureVac, Johnson & Johnson, and Oxford Biomedica texted each other. "We have to drop everything," wrote Oxford's Sarah Gilbert.[34] Because of investment from government funders, some of them military, they had a head start on some of the traditional vaccine makers, but none had yet brought a vaccine-related product to market.

As the number of deaths mounted, the World Health Organization and the Norway-based Center for Emergency Preparedness monitored the efforts under way at more than 180 institutes in China, Russia, and labs across the world. Some were synthetic biology researchers retooling to take on COVID-19. Ginkgo offered free facilities for working on a rapid COVID-19 test. As February rolled into March 2020, the virus offered synthetic biology a "transformational moment," said Jennifer Doudna, linking tiny labs and giant companies, profits and idealism in a global effort.[35]

Even before the COVID-19 vaccine race, some researchers were hailing the new science. "The synthetic biology genie was truly out of the bottle," wrote George Church, in the 2014 book he coauthored with science journalist Ed Regis, *Regenesis: How Synthetic Biology Will Reinvent Nature and Ourselves.*[36]

Some compared it to the assembly line of the automobile, or the steamship to rail connection, or the silicon revolution.

The race for a vaccine catapulted synthetic biology to the center of world events. Many of the disparate threads of a new science were joined together to confront a crisis. The thrill of synthetic biology was interdisciplinary, George Church explained to me over the phone while he was boarding a plane. "It's the quality of being curious on many topics. It features exponential progress and total engagement that makes you feel like you're making progress at a fast rate," he added without a pause. "I like seeing science have a positive effect in society."[37]

As COVID-19 spread, synthetic biology research helped produce the two most successful vaccines. It also contributed to new methods of testing and detection of new strains. Governments noticed. By September 2021, the Biden administration was proposing a $65 billion effort, in what science advisor Eric Lander called a "massive Apollo-type program" to support advanced systems to sequence pathogens found in the environment, expand lab capacity, and reduce health inequity.

Industrial synthetic biology spread to the cleaning of oil spills and sands. Its targets now numbered antibiotic resistance, safeguarding the environment, and making new medicines. As a global crisis deepened, scientists turned to viruses that attack bacteria, found in wastewater of all places. Synthetic biology became an avenue to confront an infected Earth. One potential solution came from the tiniest of killers.

10

THE KILLERS

Viruses as Healers

For them to have a little fiddle-around with these phages and to be able to make them cure something that is a huge global problem is absolutely incredible.

—Joanne Carnell-Holdaway

A s a fifteen-year-old in Faversham, England, cystic fibrosis patient Isabelle Carnell-Holdaway was dying of a drug-resistant infection after a double lung transplant. Antibiotic treatments failed. Doctors told her family to prepare for the worst. However, at the urging of Isabelle's mother, they contacted a phage research specialist.[1]

In addition to the pandemic, the world faces a crisis in antibiotic-resistant infections. Promising synthetic biology research into phages, tiny viruses that attack bacteria, may be a lifesaver. They are unseen yet live everywhere from oceans to sand dunes to our own bodies. Sporting angular, geometric heads with spindly, spiderlike legs, phage viruses resemble tiny spaceships from an alien planet. Resilient, adaptive, and numerous, these little fighters are one of the most common entities in the world that you have never heard of. Bacteriophages, or phages to the better acquainted, infect and destroy or (in keeping with the origin of their name in the Greek word *phagein* for "eat") devour bacteria.

They are most densely found in Earth's oceans and are so mind-bogglingly abundant that Yale research scientist Benjamin Chan once noted, "There's 10^2 phages on Earth. That's like 10 million times the number of stars!"[2] Although seemingly bent on world domination, these aliens are remarkably useful in combating drug-resistant diseases, making phages last resorts when other forms of medical treatment fail.

This is more their world than ours. Around 10 million trillion *trillion* phage viruses exist on Earth, outnumbering all living organisms combined. In the world's oceans, phages and bacteria are locked in an apocalyptic battle that kills, by some estimates, 40 percent of the ocean's bacteria, every day. Discovered in Britain in 1915 in water from India's river Ganges, which defeated the bacterial disease cholera, phages were used to fight Russian and German soldiers' infections during World War II but fell out of favor in the West after the triumph of antibiotics in the 1940s.[3] Then, pathogens adjusted to the antibiotic medicines. Brushed off like a dusty book, phage therapy was rediscovered in our time.

In emergency rooms, researchers mix "cocktails" of phages to fight resistant diseases. The researchers of the U.S. Navy, whose sailors often contract the antibiotic-resistant diseases of the tropics and Near East, became leaders in detecting which phages are the most effective against specific bacteria. This proved challenging because the little monsters were pathogen-specific. Like ravenous jigsaw puzzle pieces, most phages are designed to destroy a single corresponding target. They lack the blanketing death star ability of antibiotics. Another challenging quality was their location. Phages are found where bacteria thrive, so rotting fruit and vegetables, soil and sewage are prime locations.

What a scientist needed to do was dig in the muck and match a phage with its victim. But bacteria become resistant to their foes over time. Only by carefully splicing different phages together, mostly through trial and error, could researchers create medicines against the trickiest of resistant diseases.

In England, Isabelle Carnell-Holdaway's condition worsened. No antibiotics worked against her infection. At the urging of her mother, and with the help of a researcher at a London children's hospital, her doctors sent samples of her infection to University of Pittsburgh phage researcher Graham Hatfull. Hatfull was an innovative research expert on phage viruses and also ran a two-semester undergraduate program called Science Education Alliance–Phage Hunters Advancing Genomics and Evolutionary Science (SEA-PHAGES). He and his team set to work consulting his library of genomes, full genetic sequences of phages, seeking a killer equipped to take on her specific infection. The work was difficult, and time was short.[4] To understand how phages work, we begin in India.

THE PECULIAR HISTORY OF PHAGE

High in the Himalayas, at the source of the Ganges, icy, silt-laden water carries down marine fossils from the melting permafrost, as well as viruses. Locals and a British bacteriologist had observed the water's apparent antibacterial qualities but did not know what was killing the bacteria. In 1917, an eccentric French researcher named Félix d'Herelle, studying children who survived dysentery, deduced a virus was the pathogen fighter. He named it bacteriophage. A rival British microbiologist, Frederick Twort, commented on the "transmissible glassy transformations" the phages inflicted on their stricken targets, but it is d'Herelle who is credited with discovering phages' antibacterial weaponry.

By the 1920s, phages had become a life-saving medical treatment in European clinics. In the 1940s, electron microscopy confirmed d'Herelle's insight that a bacterial virus was the life saver. The effectiveness of phages was shown to be uneven and hard to manage, however, and the mass production of penicillin made it the prime antibiotic medicine of the West. Shunned but determined to continue his research, d'Herelle moved to the Soviet Union, where he helped found what was later called the Eliava Phage Institute in Tbilisi, Georgia. The capitalist world heard about phages through a novel: they were the subject of the fictional quest in Sinclair Lewis's best-selling *Arrowsmith*, a Pulitzer Prize winner about virus hunting, science rivalry, and a life of the mind.[5]

Phage therapy remained a subject of clinical studies in the United States, Sudan, Egypt, and France and became a key infection fighter in the medicine kits of German, Japanese, and Soviet soldiers during World War II. The therapy was cheap, and studies in Eastern Europe in the 1950s, and also in France, continued to develop a list of phage candidates and their targets. Its popularity in the former Eastern Bloc countries hindered its wider acceptance during the Cold War. Phages had drawbacks as well. They were hard to isolate and culture, had a narrow range of targets individually, and were less reliable than antibiotics. You could not see them except with an electron microscope. They were not as profitable either. Still, they were widely used for pure lab research for their extraordinary ability to spread their DNA. But in the 1960s, they fell into disuse as a medical treatment in the West, languishing, until the wiliness of bacteria set the stage for phages' comeback.

As stubborn bacteria like methicillin-resistant *Staphylococcus aureus* (MRSA) spread rapidly in tropical locations and through gyms and hospitals, phages sprang back to life. The U.S. Navy's research teams discovered, or rediscovered, phage therapy in an

attempt to combat the resistant bacterium *Acinetobacter bauman-
nii*, nicknamed "Iraqibacter" because it had infected so many
American sailors and soldiers in the first Iraq War. Phages were
the most diverse, abundant, least known predators on Earth. In
the 2000s, improved environmental gene sequencing methods
uncovered more and more of them, each with specific powers.
Still, few conventional medical institutions paid attention.

Then a Canadian epidemiologist named Steffanie Strathdee saved her dying husband from a vicious hospital infection.
Strathdee's husband, psychologist Tom Patterson, was dying of a
drug-resistant infection. In an Egyptian hospital, suffering from
a gallstone attack, Patterson was stricken by a killer strain of Iraqibacter. His condition deteriorated so rapidly Strathdee had him
airlifted to Germany and from there to their home in the United
States, where he was treated at the University of California, San
Diego (UCSD), hospital. But the Iraqibacter was resistant to all
available antibiotics. Soon after, Patterson fell into a deep coma.[6]
Over the internet, Strathdee and her colleagues tracked down
Texas A&M University's Rylan Young. With Lt. Cdr. Theron
Hamilton of the U.S. Navy, the two teams concocted mixtures of
phages found in barnyard sewage and ship bilges that matched
Patterson's bacterium. After Patterson miraculously recovered,
Strathdee and her colleagues went on to found the Center for
Innovative Phage Applications and Therapeutics (IPATH) at
UCSD in 2018, the first dedicated phage therapy center in North
America. Since then, it has become a resource for other desper-
ate patients and their doctors around the world. Strathdee also
chronicled her effort to save her husband in a gripping true-life
thriller, titled *The Perfect Predator*.

As fellow phage researcher Graham Hatfull looked and
looked, he settled on an unlikely candidate, or rather another
cocktail of candidates, to preserve a young girl's life.

SET TO KILL

Graham Hatfull is a soft-spoken, determined researcher and educator who had grown up near London and studied biology at the University of London before going on to the University of Edinburgh for his doctorate. He earned two prestigious postdoctoral positions, one at the Medical Research Council in England and the other at Yale in the United States. He was working on the processes by which DNA molecules exchange information when his advisor told him about the phage that infects the bacterium causing tuberculosis. The virus's ability to infiltrate DNA fascinated him.[7] Moving to the University of Pittsburgh, Hatfull won a Howard Hughes Medical Institute Award, becoming one of its "Million Dollar Professors," given the money to create a unique two-course undergraduate program in the study of phage viruses. Students would spend the first semester digging in the dirt and sampling phage DNA they found and the second semester analyzing the results. They loved it and Hatfull loved working with them.

For many phage species, the purpose of 90 percent of their genome remained unknown.[8] As one of the few labs in the West working on tuberculosis phages, Hatfull's lab had worked with numerous collaborators, including the University of KwaZulu Natal in South Africa, with the idea to attack the bacteria that cause tuberculosis and other diseases. Hatfull and University of Pittsburgh biology professor Roger Hendrix together had perfected the processes for analyzing phage viruses. They collected samples from all over the world and sequenced their genes. Hatfull's students and collaborators went into backyards to find medicines in the dirt.[9] Eventually the undergraduate program grew to include some 200 institutes around the world. As part of that effort, a phage was found in a rotting eggplant in South

Africa and was nicknamed "Muddy." Muddy, it turned out, could attack Isabelle's infection.

Until that point, phages had been used primarily as research tools by Western synthetic biology researchers. The viruses could be tapped to move genes, kill or neutralize bacteria, clean infected wounds, hack the human microbiome to treat diabetes, protect coral reefs, and diagnose and treat some types of cancer. Phages were critical in the development of CRISPR gene editing, and research like that of Northwestern's Danielle Tullman-Ercek and many others in this book began with phages. "So many synthetic biologists cut their teeth on phages," Yale specialist Paul Turner observed to me, "it is amazing." Phages would later provide critical help in the race for better COVID-19 tests and for a vaccine.[10]

These viral killers inspired a small group of devoted scientists who communicated regularly with each other. The process of testing for them in the environment was exceedingly difficult, requiring researchers to pull DNA directly from soil, waste, rotting fruit and vegetables, or sewage, and then clean and purify the sample before they could analyze it. They formed a unique community with an occasional whimsical bent, writing about their tiny subjects-of-study as if they were characters from Dr. Seuss's *Horton Hears a Who*, so that some of these scientists became known to the public.

Among them were Anca Segall and Forest Rohwer, a teaching couple at San Diego State University who once sponsored a conference with phage artwork, poetry, and music. Rohwer coauthored an illustrated book on them. "I love phages!" said the Bucharest-born, dark-haired Segall from her book-lined office.[11] "They keep a library of genes no one has seen before, rarely subjected to natural selection," she added. Her husband studied damaged coral reefs to identify phages that helped preserve those wonders from a warming, acidifying ocean. "If you save the corals," he told me, "you can save the world."

Out of everything that Hatfull and his students collected, that rotting vegetable held the key. Muddy was a virus from the bacterium *Mycobacterium smegmatis* picked to treat Carnell-Holdaway's infection. Known since 1884, *M. smegmatis* can be infected by many phages. Muddy was one of its prime antagonists. However, Muddy could not be used alone. Carnell-Holdaway's lung bacterium would soon learn how to defeat it. It was combined with two others Hatfull's lab engineered to move from a benign phase called temperate to a virulent phase called lytic. One phage, dubbed "ZoeJ," was named for the niece of the researcher who discovered it. If the combination worked, it would be the world's first synthetically engineered phage treatment.

Now the issue was keeping the three viruses viable on ice for shipment. That made a new challenge. They were racing the clock. Hatfull's team obtained Muddy and ZoeJ and ran DNA tests. They added another phage to keep Muddy from expiring. Next, they had to purify the samples, test them, ice them, and pack them for shipment. The clock was running.

At home, Isabelle was seriously weakening. She felt so tired most days she could not get out of bed. In December 2017 she was hospitalized, and by January 2018 she had been moved to the intensive care unit. Doctors advised her parents to prepare for the worst.

Hatfull's solution was packed and ready to ship, but the European Union, of which England was then a part, prohibited the import of genetically modified organisms. In a series of overnight emergency e-mails and calls with EU officials, the researchers argued that the dynamic phage duo joining Muddy featured only a deleted gene, the one that caused it to remain temperate, but not a modified gene. Finally, it was approved for importation.

On June 2, 2018, University College of London researchers raced, panting, into the hospital room. They carried a cocktail of three phages to inject intravenously. The choice of three was to increase their precision. Two weeks passed. Four. Isabelle could still barely breathe.

Then, six weeks later, the infection showed a dramatic decrease in intensity compared to before the treatment. Isabelle returned to school and started driving lessons.[12] A month later she had taken up cake baking. Her recovery was featured on National Public Radio and by the BBC. While she admitted that she found it "gross" that one of the phages came from a rotting eggplant, Isabelle and her family were grateful for the return to normalcy it allowed. Joanne, Isabelle's mother, commented, "To have a fiddle-around with these phages and to be able to make them cure something that is a huge global problem is absolutely incredible."

Researchers sought to apply the same principles to other kinds of medicines. Yale researcher Paul Turner and Berkeley's Adam Arkin founded a company called Felix Biotechnology, for d'Herelle's first name, to translate phages into clinical therapies. By 2020, eleven phage clinical trials were under way, many of them involving treatments for resistant lung infections in cystic fibrosis, and that number increased to 14 in 2022. In North America, new clinics came at phages with renewed interest. The leading U.S. labs were at UCSD, University of Pittsburgh, Texas A&M University, and Yale. The Mayo Clinic and Baylor College of Medicine were developing phage-based treatments. Much of our knowledge of phage therapy came from the Eliava Phage Institute in the East European country of Georgia and the Ludwik Hirszfeld Institute of Immunology and Experimental Therapy in Poland, where it was used as an antibacterial treatment.[13]

New start-ups joined the pursuit. Locus Biosciences focused on engineering phages for the clinic. The Maryland-based company Adaptive Phage Therapeutics, founded by a U.S. Navy researcher with a Department of Defense biosafety grant, developed phage treatments for secondary COVID-19 infections. Strathdee and Patterson invested in the latter company themselves.

The goal was to beat microbial resistance at its own game. To do so, researchers scoured the unlikeliest places for lifesavers, such as Maryland sewage in the case of Tom Patterson and the rotting vegetable from South Africa in Isabelle Carnell-Holdaway's case.[14] As phage research accelerated, however, larger-scale successes proved difficult to achieve. Many clinical trials failed through inadequate understanding of phages' complexity, and in time it was seen that Carnell-Holdaway's recovery was incomplete and her health remained precarious. For insight into how the therapy might be improved from its current state, we turn next to some other experiments that succeeded.

TURNING PHAGES

One of the first places phages played a leading role was in the labs of the early 2000s, where discoveries by Jill Banfield and others led to the breakthrough of CRISPR gene editing. Phages also played a major role in modifying RNA, for applications in drug design, nanomaterials, and the COVID-19 vaccine.

Phages taught us how little of the world we really knew. Many viruses, despite the ravages of pandemics, had made us what we are, from the gene for the placenta to the lens of the eye. If you laid out all the DNA in phages from around the world, end-to-end, they would stretch to the Perseus Cluster 300 light years away[15] Phages helped some bacteria become resistant in the first place, and the new theory proposed that phages could cripple those defenses as well.

An industry that had quietly achieved a success in phage synthetic biology was food safety. One company, which we met in chapter 2, was called Sample6. Based on research in the Jim Collins lab, the Boston-based company had a simple idea: use phage viruses to detect *Listeria* infections in food-manufacturing plants. The company built a customer base, winning a contract with Nestle, owners of Häagen-Dazs ice cream and one of the world's best-selling chocolate bars, before being sold.

At institutes like Strathdee and Patterson's IPATH and companies like Adaptive Phage Therapeutics, more clinical trials of phages were beginning to treat burn-related infections, urinary tract infections, and lung infections in cystic fibrosis patients. More than fifteen companies, including BiomX, Messaging Labs, Omnilytics and Intralytix ("lytic" for the types of phages that destroy their hosts) went to work. Maryland-based Adaptive Phage Therapeutics was developing personalized phage cocktails for multidrug-resistant diseases. Some companies, like Locus Biosciences, EnBiotix, and C3J Therapeutics, which merged with a company called AmpliPhi and became Armata, engineered phages to carry payloads for tasks like attacking resistant medical-instrument biofilms.[16] At least four companies were developing phages for waste remediation. By 2022, more than one hundred patients in Belgium and England had been treated with phages.[17] "We're a strange little family," Steffanie Strathdee told me. IPATH has been involved in the treatment of fifty-five patients, and consulted on dozens more. Yale's center has treated more than forty. At least twenty-four people received phages from the Pittsburgh group.[18]

The lab techniques work as follows. The phage knows that in order to destroy its target bacteria, it must out-strength and outnumber them. So, like the egotist it is, the phage rewires the bacteria's cellular machinery to make copies of itself. So many copies, in fact, that once all the phages are assembled, they explode outward from within the bacteria cell, shredding and killing the bacteria in the process. These little soldiers hop from bacteria to bacteria until all are destroyed. Unlike antibodies, they make sure their cleanup process is complete, digging deeper into the harmful bacteria if needed, where biomedical antibodies may abandon ship the closer to the infection they go.

The role of the scientists was to screen which phages were the most effective against specific bacteria. This was very difficult. Like ravenous jigsaw pieces, most phages are perfectly designed to destroy a single corresponding bacterium. Yet bacteria evolve quickly to resist phages, so scientists began to engineer "cocktails" to anticipate their foes' defense systems. By splicing together different strains of phages, scientists engineer medicines designed for optimum effectiveness against specific killer infections. Through this method, a patient gets a personalized phage therapy, making phage therapy one of the factors leading Western medicine in the direction of more customized, patient-specific treatment.

Phages had long been used for lab purposes, enabling researchers to transfer genes and proteins, deliver DNA, kill bacteria, and build genetic circuits. Bacterial defenses against phage infection, as we saw in chapter 4, gave the world CRISPR gene editing. Phages made the enzymes that helped to move proteins in and out of a cell and to change the amino acids on a cell membrane one at a time. They could be used as sensors, transporters, gene editors, and tumor killers. They could promote or suppress reactions inside and outside the cell. "They give us," said San Diego State phage researcher Anca Segall, "awesome tools."

Three other U.S. phage therapy centers, along with programs in Belgium, France, Sweden, Australia, and the United Kingdom, had joined the long-standing institutes in the Republic of Georgia and Poland, and IPATH in the United States, in the fight against drug-resistant pathogens. Still, the institutes were not yet harbingers of a paradigm shift in treatment. Some poorly planned and well-planned clinical trials failed. Phage effectiveness proved to be finicky and highly sensitive to individual circumstances, so much so that phage therapy defied efforts to apply it on a wide scale. "It is still something of a moving target," Graham Hatfull told a reporter in 2021, "that isn't completely pinned down."[19] Still, patients with resistant infections clamored for more phage research. "I get calls every day from people who are desperately ill," said Ry Young, professor of biochemistry at Texas A&M. "I have to tell them we're not there yet. It breaks my heart."

Indeed, once the COVID-19 pandemic spread, so much attention was paid to pathogenic viruses that people forgot that some 8 percent of the genes in the human genome are of viral origin. Viruses like phages played many important roles in our development. The most notable, perhaps, is the gene for the human placenta, which sequencing has shown was originally a gene from a virus that existed many millions of years ago. A virus gave us the ability for live, safe birth.[20]

The story of phages remained a hopeful application of pure science to the clinic. "It was Darwin on steroids," as Hatfull put it to me, meaning it was natural research combining undergraduates and faculty from around the world that led to a useful treatment for patients where antibiotics had failed.

But as a new virulence spread in the late spring and summer of 2020, most every researcher's attention was redirected. The power of biology changed the world.

A STRANGE ILLNESS

The images coming from around the world in 2020 looked like scenes from a dystopian thriller. Whole countries were placed on total lockdown. In Wuhan, China, people were allowed to leave their apartments only for an hour a week, to get groceries, and then only with passes. New York City looked deserted in gray, cold rain. Trafalgar Square was empty. The typically bustling Paris Metro looked dark and terrifying.

After the virus genome was revealed in January 2020, researchers, using algorithms, designed similar surface molecules to place on harmless viruses and trigger the body's defenses. Fueled by billions of dollars in government emergency spending, companies such as Moderna, CureVac, Johnson & Johnson, and Oxford Biomedica pursued a vaccine. They had a head start on some of the traditional vaccine makers, but none had brought a product to market.

This was synthetic biology's big moment. The World Health Organization coordinated the efforts of more than 180 institutes to create a vaccine. In China, the companies CanSinoBIO and Sinovac tapped weakened natural viruses, much as Louis Pasteur had done. In labs across the world, synthetic biology researchers retooled to take on COVID-19. Several companies and institutions, such as Jennifer Doudna's Mammoth Biosciences, Boston's Ginkgo Bioworks, Harvard and MIT's Sherlock Biosciences, and the Yale School of Public Health, turned their science platforms to making simple, cheap coronavirus tests.

In April 2020, Doudna's Mammoth Biosciences announced successful trials of a CRISPR-based at-home COVID-19 test.[21] Ginkgo Bioworks offered its facilities free of charge to COVID-19 test and antibody makers. As a key player in the pandemic battle, synthetic biology was playing many new roles.

THROUGH THE LOOKING GLASS

When COVID-19 vaccine makers ramped up their efforts in 2020, several labs turned to phage therapies to assist their research. The Russian Sputnik vaccine modified two cold viruses (adenoviruses) with phage. Johnson & Johnson's vaccine did much the same, using another common adenovirus. Adaptive Phage Therapeutics devised a phage-based coronavirus test.

From labs around the world, phage therapy offered a new vision of medicine in which viruses could be put to use. With the potential for individualized treatment, doctors could avoid the resistance-inducing overprescription of antibiotics. Spearheaded by Steffanie Strathdee, Jessica Sacher, Graham Hatfull, and others was a quest to share and improve an updated digital database of all known phages and their targets. Sacher and others annotated a database called Phage Directory. "For years, we've had different keys and different locks," said Strathdee. "But when you need a phage, often you have only a matter of days or the patient dies. You can't keep running researchers back into sewage!"[22]

The grand vision was to create an international online library, available to all but especially to medical personnel in lower-income countries where antibiotics may be difficult to obtain. Another program was to develop a Phage Exchange to share "cocktail" mixes. Then another young cystic fibrosis patient, Mallory Smith, died because she did not get a phage cocktail in time, and researchers convened at the behest of her parents. Smith's haunting autobiography, *Salt in My Soul: An Unfinished Life*, sounded a clarion call.[23] "There's got to be a better way for patients to access phage labs willing to do this work," Strathdee told me, "and for researchers to connect with one another."[24] The Phage Directory cofounders began a web newsletter, scheduled webinars, and established Phages for Global Health, a nonprofit organization to distribute phage-based medicines.

Felix, the company racing to adapt phages for treatments, felt it was close to being ready. "We think we can be in clinical trials in a matter of two years," Turner predicted to me in 2021.[25]

An outspoken voice making patients' case for phage treatments was advocate and consultant Ella Balasa, who was meeting the challenge of her own chronic lung infection. As a medical student in Richmond, Virginia, she had to leave school for a time to deal with a resistant infection much like Carnell-Holdaway's. She was the one who brought phage to the attention of her doctor. "Phage treatment saved my life," Balasa told me.[26] Balasa has testified in hearings and amassed a social media following to promote the work of new phage researchers. In the meantime, Carnell Holdaway's health declined as she struggled with her resistant lung infection.

In Boston, the small start-up PhagePro developed a three-phage vaccine to prevent the spread of cholera in impoverished households, winning a 2021 grant to move to clinical trials.[27] As the field gained interest, Steffanie Strathdee pointed to the massive COVID-19 vaccine investment as a model for what it would take to make the therapy more widely available.

In the United Kingdom, Carnell-Holdaway's phage treatment no longer held her infection in check. In the spring of 2022, she passed away. "The type of infections currently targeted by phages are often the really rough ones," observed Graham Hatfull, noting the young girl's courage in the face of an intractable infection. There is a "real need to progress to progress to formal clinical trials where we can really learn about safety and efficacy.[28] It is not helpful to think of phages as 'miracle drugs.'"

Other efforts were under way to improve on phages' potential power through improvement by gene editing, as Strathdee and Hatfull highlighted in a January 2023 *Cell* review article.[29] At the same time, researchers sought other ways to exploit viruses to deliver medicines, and a promising new Ebola vaccine used a synthetically engineered virus. Still, every step forward prompted the microbes to push back, and phages remained in some ways a mystery. "You can get a readout of (phage) genes," said a phage specialist Karen Maxwell of the University of Toronto in *Nature Biotechnology*, "but unfortunately . . . (often we have) no idea what they're doing." For that, researchers envisioned a plan to augment, not replace, antibiotics. "Not phage *or* antibiotics, but phage *and* antibiotics," became a mantra.

Then the COVID-19 vaccine race moved synthetic biology to the fore of global science, world health, and public and venture investment. Vaccine development, testing, and treatment efforts stumbled and lurched forward simultaneously. We cover the genetic engineering of vaccine development in chapter 11 and of vaccine production in chapter 12, the ethics of synthetic biology in chapter 13, and synthetic biology in space in chapter 14. Chapter 15 answers the question, Will this new science make for an industrial revolution? Part III describes what happened, and why.

III

BIOINDUSTRIAL REVOLUTION

11

RACE TO A VACCINE

I never doubted it would work.
—Katalin Karikó

n 2019 in a small biotech company in Cambridge, Massachusetts, scientists were struggling. The company, Moderna, had been working on the tiny, single-stranded instruction transmitter of cells, messenger RNA (mRNA), to attack diseases caused by defective proteins, such as a condition that causes jaundice.[1] Although the company had raised a billion dollars, its business strategy had not worked. Under its CEO, the French-born Stéphane Bancel, the company pivoted to vaccines and took a special interest in the severe respiratory viruses emerging in pockets around the world. The plan, also embraced by companies like BioNTech of Mainz, Germany, was to create vaccines by programming mRNA with new instructions to create virus proteins that would stimulate the body's defenses. The use of mRNA, life's universal software, promised greater speed and adaptability than what was possible in traditional vaccine development. In an interview in *Science*, Moderna's president,

Stephen Hoge, explained that if they cracked the rules of mRNA, "essentially the entire kingdom of life is available for you to play with."[2]

Investors such as Alexion Pharmaceuticals poured hundreds of millions of dollars into the company, but Moderna's vaccines had failed to make it to market. A partnership with pharmaceutical giant AstraZeneca that focused on the immune system and cancer had not panned out. Moderna had no sales revenue. Then the coronavirus broke out in Wuhan.

The first step to making a vaccine was choosing how to attack the virus. Use of mRNA requires editing the genetic messenger to direct the patient's cells to produce weaker versions of the virus's dangerous spike proteins. The next step is to test versions in mice to ensure that the protein sparks the immune system to create antibodies. To do that, samples of mRNA are manufactured through lab chemistry. The third step is to optimize one best version, then manufacture human test vaccines and make sure they are safe, setting up clinical trials to see if they work. The final step entails the massive logistics of fast, reliable, and adequate production and distribution. Complicating a mRNA vaccine is its fragility, requiring supercold storage, leaving many experts to question its reliability for world distribution.

Normally, it took ten to fifteen years to make a new vaccine, as in the cases of Jonas Salk's and Albert Sabin's breakthrough polio vaccines in the 1950s. Until 2020, the previous fastest development had been the nearly five years it took to develop a mumps vaccine, which became available in 1967.[3] But with global business, travel, trade, and almost everything else at a standstill, and the world death toll soaring, no one had that time.

BABY STEPS

After a Chinese scientist and an Australian scientist together revealed the severe acute respiratory syndrome coronavirus 2 (SARS-CoV-2) genome, listing the sequence of the RNA virus in January 2020, researchers around the world raced to understand how best to attack it.[4] Its name reflected its crown-like appearance under the microscope.[5] Thanks to the work of the preceding years, the BioNTech and Moderna researchers quickly identified target sequences for making mRNA that would muster the human immune system to develop its defenses. The body would be the manufacturer.[6]

February 2020 marked the first published U.S. death from COVID-19, although others most likely occurred earlier.[7] SARS-CoV-2 presented several challenges. It was a larger virus and more transmissible than the earlier SARS viruses, and it contained "bells and whistles" such as an enzyme deactivator of the "off" switch that would normally stop the human system from overreacting.[8] Thus, many patients experienced a gyrating immune reaction that almost defied treatment.

The mRNA-focused companies were not the only ones scrambling for a vaccine. Others were using the virus's sequence to produce conventional weakened versions of the virus, still others to make a cutting-edge DNA vaccine, and at least three groups were trying to modify a chimpanzee cold virus to induce the human immune system to make the antibodies to protect itself.[9]

But synthetic biology in the lab could do more with a killer than speed the development of a vaccine. It could help make tests and treatments. A critical problem of the U.S. COVID-19 response was the lack of reliable tests. Without tests, thousands of people waited hours in lines or suffered at home as the pandemic kept spreading. Synthetic biology contributed to the development of cheap, reliable at-home tests, and it worked on antibodies and other new ways of attacking the disease.[10]

Still, the Holy Grail was a vaccine. The mRNA vaccinemakers made snippets of DNA sequence, which in turn made the mRNA they would use to engineer model vaccines to test in mice. At Moderna and BioNTech, researchers transferred the newly minted DNA to *E. coli*, which dutifully divided, their offspring featuring the modified DNA. The extra DNA then produced mRNA for tests in animals.

One advantage to companies like Moderna, Oxford, and BioNTech was their previous experience. The infrastructure and techniques had been developed and tested. Oxford had run clinical trials on a vaccine for the Middle East respiratory syndrome (MERS) coronavirus and had a factory ready to make doses. It started screening volunteers for COVID-19 vaccine safety trials and discussing manufacturing and shipping with AstraZeneca. Moderna was at much the same point because of its Zika vaccine development. BioNTech had been working with pharmaceutical giant Pfizer on a flu vaccine and, in March 2020, agreed to collaborate on a coronavirus vaccine.[11] That groundwork was also a testament to early funding decisions by some government agencies that supported mRNA vaccine development. "We got lucky that mRNA vaccines worked," said former FDA commissioner Scott Gottlieb. But it was not luck alone.[12]

Time was so short that researchers had to run each stage—build, produce, test safety, screen efficacy—simultaneously rather than in succession as they normally would. The presence of numerous critics, ready to assail most any mistake, added further pressure. There was a huge amount of data to analyze.

Failure on any of those fronts could derail the effort. The issues of manufacturing, equal access, and difficulties of storage had to be confronted. "This pandemic was going to define how people thought about biotechnology for the rest of their lives," commented Berkeley's Jennifer Doudna.

Dozens of other competing institutes raced the mRNA modifiers to come up with a vaccine through more tried-and-true processes. In Russia, the Gamaleya National Center of Epidemiology turned to the adenovirus. Many utilized synthetic biology to some extent, but not in the new way of Pfizer and Moderna.

A major component in the race was an unstated competition. "We're not in a race against other humans," commented Oxford's Sarah Gilbert. "We're in a race against a virus."[13] Perhaps. As the two mRNA vaccine teams went after the virus with all the tools of synthetic biology, the promise of a new science was put to the test.

FIRST MONTHS

Already in May 2020, BioNTech was injecting the first U.S. patients in a combined Phase I and II safety trial of its vaccine and had concluded its agreement with Pfizer to manufacture it in bulk. In July, Pfizer signed a deal with the U.S. government to provide up to 600 million doses. The preceding years of work enabled BioNTech to move at a speed and scale never seen before. BioNTech CEO and cofounder Ugur Sahin announced, "We have been able to leverage more than a decade of experience in developing our mRNA platforms."[14]

Back at Moderna's lab in Cambridge, Massachusetts, the need for a coronavirus vaccine had outweighed concerns about

Moderna's somewhat barren earlier pipeline. Its stock price shot up 200 percent from January to June 2020. Moderna's work to develop the mRNA-1273 vaccine relied on genetically engineering DNA to make the mRNA it needed. Cash for the project came from the U.S. government's Operation Warp Speed, and the Biomedical Advanced Research and Development Authority (BARDA) awarded the company a $483 million grant. Moderna signed a deal with manufacturer Lonza to make from 500 million to 1 billion doses.[15]

On May 18, 2020, Moderna issued a press release announcing a successful Phase I safety trial. However, because the results were published in a release instead of in a peer-reviewed journal, data and details were limited. Still, that month, the company launched Phase II clinical trials to prove the vaccine worked and proceeded with plans for Phase III trials on 30,000 individuals to start at 100 sites in the summer.[16] On Monday, July 27, 2020, volunteer Melissa Harting of New York became the first participant to receive a dose.

While Moderna and Pfizer vaccine-makers confronted the extremely cold temperature required to keep their vaccines viable, CureVac, a smaller German company, was making an easier-to-store, more easily tolerated, lower-dose RNA-derived vaccine. However, CureVac lagged in its progress, despite getting government funding on a par with Moderna and BioNTech.[17] Still, hamster experiments proved its vaccine worked fairly well. By June 2020, the German government had invested some $360 million, and by the end of the year the company began final trials. Those, however, proved a disappointment, perhaps due to the lower dose or increasing number of COVID-19 variants. In June 2021, CureVac's vaccine was found to be only 47 percent effective.[18]

Elsewhere, other technologies were employed, and the race continued.

"THE ONLY OPTION FOR US
TO GET IT ON TIME"

Johnson & Johnson was first among well-established companies to pursue a coronavirus vaccine.[19] Its Janssen Pharmaceutical subsidiary collaborated with a team from Boston-based Beth Israel Deaconess Medical Center to develop a medicine for clinical trials. The company received $456 million in government funding. Nor was it starting from scratch. When the Zika epidemic emerged in 2015, Johnson & Johnson researchers produced a vaccine using a cold virus studded with faux Zika proteins. The Johnson & Johnson coronavirus vaccine was similarly based on the pandemic virus's genetic instructions for making the spike protein. But the Johnson & Johnson vaccines used double-stranded DNA, not the single-stranded RNA of Moderna and BioNTech, to deliver the instructions.[20] Scientists added the spike protein gene to a common cold virus known as adenovirus 26. The vaccine created antibody responses in primates, with human clinical trials commencing in late July 2020 in the United States and Belgium.

Money poured in. In August 2020, the federal government agreed to pay Johnson & Johnson $1 billion for 100 million doses. In September, the Johnson & Johnson vaccine went to a clinical trial, but on October 12 the company paused it to investigate one adverse reaction. After falling behind its original production schedule, in January 2021 the company released its results, showing a 72 percent overall success rate in the United States and 64 percent in South Africa, where the highly contagious B.1.351 variant was driving cases in 2020. The vaccine also showed efficacy against severe forms of COVID-19. In February 2021, the Food and Drug Administration approved the vaccine, first for emergency and then for general administration.

Such a rapid turnaround was only possible because Johnson & Johnson committed to production of a billion doses globally even as it was still testing the vaccine. The company conducted Phase III efficacy studies in North America and Europe, with plans to branch into countries such as Brazil and South Africa. Paul Stoffels, Johnson & Johnson's chief scientific officer, told Reuters the company had to start ramping up manufacturing capacity immediately, even before it had a clear signal that its candidate worked. "That is the only option for us to get it on time," Stoffels said. The company had a manufacturing plant in the Netherlands that could make 300 million doses, but more capacity was needed.[21] Ultimately, the maker of everything medical from Band-Aids to baby powder deployed a weapon twenty years in the making.

IN TIANJIN

Some 6,000 miles away, the Chinese company CanSino Biologics (CanSinoBIO) was accelerating its vaccine trials with equal frenzy. Like Moderna, CanSinoBIO was a relatively new start-up, founded in 2009 in Tianjin, China. Tianjin was a sprawling manufacturing center located south of Beijing. CanSinoBIO's technology was based on four research platforms: adenovirus-based viral vector vaccines (developing vaccines from the cold-causing adenovirus), conjugate vaccine technology (a type of vaccine utilizing a weak antigen and a strong antigen), protein-based vaccines, and cell culture formulation. Two late-stage meningitis vaccines were in development, and one Ebola vaccine had received new drug approval in China in

October 2017.[22] However, the company had experienced a $20 million loss in 2019.[23]

In May 2020, CanSinoBIO had made global headlines when it became the first company to publish its findings from Phase I vaccine safety trials in a peer-reviewed medical journal. The CanSinoBIO vaccine was also an adenovirus-vectored drug using SARS-CoV-2 genetic material, similar to Johnson & Johnson's vaccine. In the study, 108 participants aged 18 to 60 received high, medium, and low doses of the vaccine. While T cell and immune responses were promising, some 80 percent of participants reported fairly mild side effects, including fever. Still, by being transparent about the data by publishing the study in *The Lancet*, the company set a precedent for researchers worldwide.[24]

On July 20, 2020, CanSinoBIO published the results of its second trial of 603 patients. Phase II was conducted in the same way as the first, with participants given one of three different doses of the vaccine undisclosed to either administrator or patient to control for any placebo effect. There was a marked difference in side effects between high and low doses. High doses of the vaccine produced severe side effects in 9 percent of volunteers, while lower doses only produced severe effects in 1 percent, compelling researchers to scrap the highest dose. Still, the vaccine was sufficiently successful to be approved by the Chinese government in late 2020 for emergency use for soldiers, a global first in the vaccine race for approval.

In total, the lesson of the vaccine race was that if one invested billions of dollars, one may realize immediate, sweeping benefits from pure research such as what had been happening in synthetic biology. Meanwhile, a huge question hung over the pandemic. Where had the coronavirus originated?

THE QUESTION OF ORIGIN

The international debate over the origin of the pandemic virus that killed about 6.9 million human beings invoked lab practices, biosecurity, "gain-of-function" lab manipulation of viruses, and the highest political stakes." Several books have summarized what is known to date. None have solved the most profound mystery of our time. However a brief summary is critical in judging when and where lab biology has been either a savior or a potential cause of disaster.

Viruses are studied and altered most every day in hundreds of labs around the world. Thanks to the work of Chinese and other researchers, we know that the SARS virus, the predecessor of COVID-19's SARS-CoV-2, originated from bats in caves in southern China, having passed through a mammal in the live markets of traditional Chinese medicine (TCM) in 2003, probably palm civets in Shenzhen, where they were sold as food. SARS is a zoonotic disease, meaning it originated in animals, as so many other viruses have. The mammals are bitten by the bats and then captured for food. Animal handlers were among the first infected by SARS.[25] The coauthor of the 2017 *PLoS Pathogens* paper that solved the mystery of the SARS virus was Wuhan virologist Shi Zheng-Li, who would come under criticism by supporters of the lab leak theory in the case of COVID-19.[26]

When, on January 11, 2020, the SARS-CoV-2 genome was posted online at some risk by seasoned virus hunters Edward Holmes and Zhang Yongzhen, researchers noted something unusual.[27] The spike proteins on its shell grabbed more tightly onto human cell receptor proteins than to receptor proteins of any other species. The spike featured a furin cleavage site, a unique insertion of four amino acids on the protein not present in other related coronaviruses. It was almost as if that part of SARS-CoV-2 had been designed to attack humans in particular. As Australian virologist Nikolai Petrovsky summed it up to the magazine *Undark*, "That's weird. Holy shit."[28]

To determine the origin of the virus, the World Health Organization (WHO) sent a delegation to Beijing in January 2020, but the team did not receive full cooperation from the Chinese government.[29] Investigators were accompanied at all times by government officials and were not allowed access to the Wuhan Institute of Virology where three workers had been hospitalized for flu-like symptoms the preceding November.[30] Nevertheless, researchers suspected that the origin of SARS-CoV-2 was similar to that of the SARS virus, potentially through the Huanan Seafood Wholesale Market where many early cases clustered. The WHO team suggested that the virus originated in cave bats and was carried by an intermediate host animal to humans.

The seafood market was disinfected and closed by the start of 2020. The trouble was, of the initial DNA tests of the Huanan seafood market animals, it appeared that none showed infection by SARS-CoV-2.[31] Pangolins, anteater-like mammals also sold as food in Chinese wet markets, were named as the prime suspects for transmission by researchers at the Agricultural University of Guangzhou.[32] However, in the 2020 paper proposing pangolins as the intermediate host, only two out of one hundred pangolins tested positive for SARS-CoV-2, and these for a weak form of the virus that may have come from infection by a worker.[33] However, a newly released, unpublished report suggested that raccoon dogs in the market may have been infected with SARS-CoV-2.

The Wuhan Institute of Virology was both a leading virology research center and a source of quality control questions. It opened in 1956 as a microbiology facility focused on soil pathogens to help farmers. Gradually it switched to virus research, and some of the world's top researchers trained its staff. The Wuhan virologist Shi Zheng-Li was an expert on bat coronaviruses, many of which were not dangerous to humans. The Wuhan Institute was the world's leading laboratory in coronavirus study, with precautions against leaks exceeding those of western labs, reported Royal Melbourne Hospital virologist Danielle Anderson, the last western scientist to work there. (Anderson's time there ended in November 2019). Still, in 2017, Wuhan lab researchers themselves had complained.[34] In 2017, lab researchers themselves complained of a lack of appropriate training, prompting two visits by U.S. scientists who warned the State Department of some staffing issues. The Wuhan Institute of Virology had only recently received the biosafety level 4 (BLS4) clearance, the highest such clearance, in 2018. It housed the world's largest collection of coronaviruses.

In February, a letter signed by twenty-seven scientists, published in the British medical journal *The Lancet*, proclaimed that the virus must have had a natural origin. But the almost immediate blanket conclusion only deepened suspicions. A year later, a Wuhan site investigation by WHO-commissioned scientists was also inconclusive. It concluded, however, that a lab leak was "extremely unlikely."

Another problem was the Chinese government's failure to disclose that a SARS-like coronavirus had been collected in an abandoned mine cave in Yunnan, near the Laotian border, sickening six miners back in 2012, and killing three.[35] The virus was studied at the Wuhan Institute, but is not similar enough to the pandemic virus to be called its progenitor.[36] Finally, in 2021, President Biden authorized a bipartisan U.S. panel to address the many questions surrounding the origin of COVID-19. Its final report was inconclusive but one "IC element," intelligence community element, assessed with "moderate confidence" that a lab leak was the pandemic's origin.[37] That was the FBI.

A provocative issue was to what extent the National Institute of Allergy and Infectious Diseases (NIAID), headed by Anthony Fauci, had indirectly funded the Wuhan research.[38] From a $3.7 million NIAID grant to a New York–based nonprofit, the EcoHealth Alliance, to prevent pandemics, $600,000 was to be provided by EcoHealth to the Wuhan Institute of Virology to study coronaviruses in a collaboration planned for the period 2016–2024, at some $76,000 per year. The NIAID grant forbade gain-of-function viral study, and required the reporting of any adverse lab results, under a later rule called P3CO. A study of the full 528-page description of the grant, released in September 2021, suggested the China lab may have altered SARS-like viruses to study their pathogenicity.[39] Few were accusing the lab of intentionally leaking a killer, but there was suspicion a stored virus could have escaped by accident.[40]

In support of those who blamed an accidental lab leak, in September 2021 it was revealed that the EcoHealth Alliance had applied for a DARPA grant to research the insertion of furin cleavage sites into SARS viruses, to be completed at the Wuhan Institute of Virology—exactly what the SARS-CoV-2 virus proved to be. The grant was rejected, and EcoHealth claimed none of the research was pursued. With all of the accusations, the EcoHealth Alliance grant was canceled in April 2020. Eco-Health was one of eleven institutes, however, awarded five-year NIH grants in August of that same year.[41]

Those arguing for a natural origin could point to the fact that most every pandemic in history has been accused of coming from human origin. The Russian government claimed Zika and Ebola viruses were the results of American clandestine programs. Some thought the same of AIDS. More to the point, a Pasteur

Institute team report published in *Nature* in 2022 revealed that the team had found bats in several caves of northern Laos and southern China with coronaviruses that could infect humans. The discovery suggested that coronaviruses emerge frequently in bats, often without being identified, and might not need an intermediate host to spread to humans. The causes for transmission to humans might be our encroachment into remote habitats, in part in search of guano as fertilizer.[42] A University of Hawaii study in 2021 suggested that a warming climate was fueling an increase in bat species in southern China, and that outbreaks of coronaviruses are related to increasing numbers of bats in a region.[43]

A 2022 study published in *Science* concluded that the Huanan Seafood Wholesale Market in Wuhan was the early epicenter of COVID-19.[44] But the passage of time and suppression of data make it hard to trace the original chain of events. Answering the question of COVID-19's origin remains critical however, if public health officials are to plan for the next pandemic. The risk remains high, as human population growth brings us into closer contact with more wild animals.

Biosecurity would be a critical foundation of a bioeconomy. "We need a comprehensive plan so that our lack of preparedness never happens again," said former FDA commissioner Scott Gottlieb. Wuhan virologist Shi Zheng-Li agreed. "This pandemic has made me realize the importance of our work," she wrote in a reply to *Science* magazine. "With global environmental change and the expansion of human activity, the risk of infection continues to increase."[45]

Professor Shi vigorously defended herself, and then mostly stopped talking in public.[46] She identified bats as the likely origin of SARS-CoV-2 and noted that "the places where big emerging diseases break out usually are not their source of origin."[47]

She cited the high rates of virus antibodies found in Wuhan's house and stray cats, proposing that the virus was spread from humans to cats. The bottom line, she argued, was that the Wuhan Institute of Virology received its first samples of the virus back on December 30, 2019, when it was labeled "pneumonia with unknown etiology."[48] She denied doing any unpublished gain-of-function experiments with coronaviruses and ruled out the possibility of an accidental release from her lab.

If one good result comes from the pandemic, it could be the establishment of globally enforceable gain-of-function research guidelines and coordinated public health responses for future pandemics. For the moment, however, certainty seems unlikely, and the origin question remains to be answered.

"THE GREATEST SCIENCE EXPERIMENT IN VACCINOLOGY THAT'S EVER BEEN DONE"

As 2020 slid from summer to fall, international air travel slowed to one-third of its pre-pandemic level. Vacations became regional or national, and the sales numbers for recreational vehicles, national park passes, and campground reservations skyrocketed. That summer, my family was fortunate to be seated around a fire outside Glacier National Park, awaiting the rare appearance of the comet NEOWISE. The falling comet slashed the black horizon behind the Rocky Mountains. As we played a guessing game about quotes from our favorite movies, the Milky Way materialized like a giant shroud.

Around the country, people were protesting the killing of George Floyd, state mask mandates, layoffs, and quarantine restrictions. Many waited in long unemployment lines. Many

Americans of Asian descent were harassed, including one of my students, whose Michigan high school locker was covered with COVID-19 stickers.

With tens of billions of dollars being invested in prevention and treatment around the world, the COVID-19 vaccine race became synthetic biology's moment. Those words of Jennifer Doudna were echoed by BioNTech CEO Ugur Sahin and Scripps Research Institute virologist Andrew Ward, who called the research programs "the greatest science experiment in vaccinology that's ever been done." Finally, after twenty years of promise and potential, a new science was helping to solve a world crisis.

Synthetic biology played its role in the pandemic response in three ways: through the development of RNA- and DNA-based vaccines; through diagnostics; and through therapies. In diagnostics, several synthetic biology at-home tests were being developed, some saliva-based and others using the traditional nasal swab. Some detected antibodies and others viral proteins. One used crowd-sourced detection systems developed at the University of Washington. With COVID-19 tests in short supply at the pandemic's outset, several synthetic biology companies stepped up their efforts to create tests to detect antibodies. Two such synthetic biology companies, Sherlock Biosciences in Cambridge, Massachusetts, and Mammoth Biosciences in Brisbane, California, used CRISPR to engineer microbes to detect the antibodies. "It felt like being in the World Series," said Christine Coticchia, Sherlock's former principal scientist, in a radio interview.

Many COVID-19 test kits required tedious liquid transfer by handheld pipette or small glass tube. From New York, the small company Opentrons made inexpensive robots called OT2 to perform such liquid transfers. During the pandemic, its robot sales skyrocketed to include some forty countries as well as its home city when it won the New York contract to make the robots for city tests.[49] Use of the robots drastically reduced the wait time for test results. Will Canine, a cofounder of Opentrons and former protester in the Occupy Wall Street movement, saw his company's valuation leap in one year from $90 million to $1.8 billion.

Ginkgo Bioworks also made COVID testing kits and shipped them free to public schools across several states, while Northwestern University and Yale School of Medicine researchers turned their laboratory platforms to making alternative, inexpensive coronavirus tests. The Ginkgo effort was a partnership with the company Concentric to collectively pool tests from several patients at once, lowering the cost of COVID-19 testing, securing access to tens of millions of rapid antigen, or surface proteins, tests a month. Northwestern's Center for Synthetic Biology offered two different kinds of inexpensive tests using cell-free technology, techniques that did not require living cells. The company Distributed Bio partnered with the World Health Organization and the U.S. military to create a universal vaccine.[50] In Illinois, Argonne National Laboratory enlisted its artificial intelligence algorithms to track the virus's evolution, to predict mutations, and to anticipate the next new virus outbreak.[51]

Other companies created versions of the virus protein to help test therapies and to make synthetic biology COVID-19 therapeutics themselves. These included antibodies, weakened proteins, and even synthetic steroids. Twist Bioscience in San Francisco, for instance, used its unique silicon-based DNA writing platform for viral detection and analysis, and its clonal genes allowed for quick development of vaccines from gene sequences, which relieved lab scientists from having to handle dangerous pathogens. The laboratory of Jim Collins, now at MIT, worked on a mask that could turn blue if the virus was detected in the wearer's breath.[52]

Newer synthetic biology companies focused on COVID-19 as well. In California, Antheia worked on therapeutics. Academic researchers like Yale's Paul Turner made COVID-19 a teaching tool for viral evolution. As winter 2020 rolled into the new year, vaccine makers turned to manufacturing their medicines in bulk. In the turn to global manufacturing, synthetic biology offered a vision of what future bioindustry might look like, which is where we turn next.

12

GLOBAL PRODUCTION
Perils and Profits of a New Science

A revolution was made by people talking . . . about something
theoretical, something they hoped *would exist. It was a dream.*
—Malcolm Gladwell, *The Bomber Mafia*

I t was a beautiful morning, and I was sitting at my usual
kitchen spot at the granite island, studying my laptop, click-
ing for what seemed the hundredth time through the websites of Zocdoc, CVS, Walgreens, Jewel-Osco, Publix, Good
RX, App Doctor, and every other site I could think of, looking
for a vaccine appointment. Outside, the sun baked down on
the still water of the Gulf of Mexico. All I could see was "none
available."

In the spring of 2021, Pfizer and Moderna competed to
produce their vaccines for millions of people. It was one thing
to make mRNA vaccines in small doses in the lab, but quite
another to build facilities to make enough to vaccinate a country
or a continent. The equipment and personnel costs were tremendous. The fragile, organic ingredients interacted differently in a
small fermenter than in a gigantic 20,000-liter bioreactor, where
the pressure built up higher and the precious product had to be

separated from hundreds of gallons of broth.[1] Even differences in tap water quality affected the outcomes. Pfizer had promised 100 million doses by 2021 but fell short by half, with an assurance of more to come with money from the U.S. government.

The vaccine makers were solving problems that faced synthetic biology as a whole: how to turn lab successes into a business. By mechanizing the production of organic material at unprecedented speed and scale, the efforts showcased the required logistics of industrial synthetic biology. What was the proper sequence and mechanism for feeding gigantic fermenters? Should workers perform tasks by hand or could processes be automated? Can yeast or bacteria be induced to ferment 24 hours a day? In the case of the Pfizer vaccine, how could the product be kept at temperatures as low as 70 degrees below zero Fahrenheit? The solutions required a combination of chemical and mechanical engineering and sophisticated quality control. Manufacturing needed thousands of skilled workers toiling extra hours, wearing masks and bulky protective gear amidst a global shutdown.

For many people, the lockdown rolled past in Zoom meetings, sleepy classes, home schooling, new pets, and work from home. For others, it meant furloughs, layoffs, creditor calls, and anxiously awaiting a stimulus check. Many of us felt we had been waiting for years for the vaccine. I was one year shy of the age requirement for Priority One inoculation status in my native state of Illinois. I scrolled impatiently, not daring to lie about my job status and cut in line only to be humiliated on social media: "Professor caught lying on vaccine questionnaire!" It sometimes felt like the worst part of the vaccine effort was seeing celebrities pictured with bared biceps.

By January 2021, more than 2 million people were dead worldwide, with 300,000 or so gone in the United States.[2]

Twelve months later, 5.9 million people had lost their lives.[3] The pandemic ebbed and surged. A series of network news programs airing at the one- and two-year anniversaries of the pandemic doled out the results. Hundreds of millions sick, tens of millions unemployed, variants emerging. People coped by writing haikus (three-line poems of meditation) and scheduling Zoom sessions with therapists, and those were people with money. Compared with the 17 to 100 million deaths that occurred during the 1918 influenza pandemic or with the Plague ("Black Death") of the 1340s, which claimed one-third to one-half of Europe, COVID-19 was less lethal but more draining because of its endless social media fractures and accusations.[4] My wait for a vaccine seemed to drag on.

There were hopeful signs. By 2022, vaccine availability was much less of a problem than it had been. Why? The answer is in the ways the vaccine makers achieved mass-production levels. Things did not always go smoothly, but the surge of production demonstrated some of the necessary technologies, capacity, and capability to make synthetic biology a force in manufacturing.

"A SECOND MIRACLE . . . OF MANUFACTURING"

"Necessity is the mother of invention" is a saying that arises during a crisis. What better time than a crisis for revolutionary technology to take hold, when its appearance could ignite widespread, immediate improvement in people's lives? In the United States, vaccine makers were reworking that common adage to another philosophy that instead stated, necessity makes a child of innovation. "Making a vaccine is quantifiably one of the hardest things human beings try to do" said Regeneron cofounder

George Yancopoulos in the *New York Times Magazine* in June 2020.[5] The companies struggled at first.

The first COVID-19 vaccine approved for emergency use in the United States was Pfizer's, in December 2020, with Moderna winning approval shortly thereafter.[6] Pfizer had only signed its production deal with BioNTech some eight months earlier. With the approval, the partnership had to speed into industrial production and its huge challenges. Pfizer faced the added complication of having turned down U.S. government funding initially. Pfizer facilities in Missouri, Massachusetts, and Michigan, along with its base in Mainz, Germany, were retooled to meet the demand. The effort's meaning resonated with employees. "Rarely do you see something you work . . . have an effect outside of the lab" said one research manager to the *New York Times*. "This was coming home and seeing ten headlines a day."[7]

Pfizer converted its plant in Chesterfield, Missouri, to produce the DNA templates for the mRNA vaccines, combining tiny loops called plasmids containing the instructions to make coronavirus spike proteins.[8] This process thus began with premade snippets of genetic material. The DNA was manufactured in enormous quantities in fermenting vats of *E. coli* bacteria modified to take up the premade genetic material. The fermenters at the sleek, 295,000 square-foot facility resembled those of an industrial brewery. As these *E. coli* reproduced rapidly, each new generation produced more and more of the desired DNA template.

Scientists recovered the bacterial DNA by shearing the cells, dissolving their components, then distilling the extracted DNA by chemical methods. The DNA was employed to produce the vaccine's mRNA. This was done by cutting the bacteria's DNA with enzymes, making small linear strips, purifying them, and storing the DNA in one-liter bottles, each containing enough genetic material to make 1.5 million doses. This was very similar to technologies used by biotechnology companies in the 1990s, except at vastly larger and more sophisticated scales. The bottles were frozen and shipped to one of two Pfizer plants, either in Andover, Massachusetts, or in Mainz, Germany, where the contents were translated into fragile mRNA, tested, frozen, and shipped to Kalamazoo, Michigan, to be thawed, mixed with water, and encased in the vital fatty lipids that held the vaccine, then frozen at 70 degrees below zero Fahrenheit to be shipped to patients.

The Kalamazoo operation was arguably the most critical. If the plant could not push production, then all the grueling twenty-four-hours-a-day, seven-days-a-week scheduling would be for nothing. The midwestern town was home to one of America's first drug manufacturers, Upjohn. Pfizer had acquired Upjohn's sprawling plant, built in 1948 as a producer of steroids and snakebite antivenom medicine. The building was nearly a quarter-mile long to house the various machines and store their supplies and products.[9] It featured four tall white towers resembling grain silos, for storing the dry ice needed to encase the vaccine for shipment. In September 2020, the Kalamazoo plant made its first test run, pumping precious messenger RNA to be encased in lipid, or oil, membranes.[10] Gigantic sub-zero freezers were ready to preserve the vaccine vials. Ductwork beneath the ground sucked out the hot air created by the freezers.

The first test failed. Engineers traced the problem to a faulty filtering membrane that leaked the precious lipid particles. They corrected the membranes for a similar test in Pfizer's Puurs, Belgium, manufacturing plant, and this succeeded.[11] But the filters remained a weak point even as clinical trials showed how well the vaccine worked. The clinical successes put more pressure on the Kalamazoo workers, identifiable by their bright yellow shirts as they packed the bottles in the dry ice, preparing to ship them the instant the FDA granted emergency approval. By the time the approval arrived in early December, Pfizer had cut its 2020 production prediction in half, from 100 million to 50 million doses. Needing money, the company sought U.S. Defense Production Act authorization to purchase packing supplies, and the new Biden administration agreed.[12] The new president called it "a second miracle, the miracle of manufacturing."[13]

Millions of vaccines shipped in outsized containers. The complicated handling required high-tech equipment and large-scale organization. Some journalists compared the scale to that of the Apollo space program, but this much bigger operation more personally affected the lives of everyday people. The newly developed expertise and processes, equipment and training, could prepare the groundwork for what could be a bioindustrial revolution, using the raw materials of nature for sustainable medical, material, food, and clothing production.

TREATMENTS IN THE PIPELINE

By the end of 2021, synthetic biology expanded with more than $4 trillion invested worldwide. Ginkgo Bioworks catapulted to Wall Street and was valued at $15 billion. The top companies included familiar names such as Ginkgo, Amyris, Bolt Threads, Antheia, LanzaTech, Pivot Bio, and Impossible Foods, along with Ames, Iowa–based Renewable Energy Group or Eddyville, Iowa–based Cargill. *Iron Man* actor Robert Downey Jr. invested in synthetic meat makers. In Skokie, Illinois, LanzaTech partnered with Lululemon to make yoga pants out of steel-mill carbon exhaust, which it was using also for detergents, and with Zara to make a black cocktail dress from the same source.

Next to tackle was how to distribute synthetic biology–created COVID-19 treatments. Ginkgo Bioworks, Twist Bioscience, and Antheia were among those providing crucial services for free or at reduced rates. Ginkgo committed much of its technology to better testing and offered its test for COVID-19 free of charge to schools. The company also offered its automated tools to support COVID-19 surveillance, tracing, and manufacturing.[14] Twist Bioscience likewise offered its silicon gene manufacturing tools for viral detection and as prized RNA controls.[15] Integrated DNA Technologies, the thirty-year-old DNA synthesis company, provided coronavirus samples to qualifying labs. ATUM, another DNA synthesis company, lowered costs and increased the accuracy and efficiency of gene manufacturing to order.

Some critics questioned the wisdom of making potentially dangerous samples of SARS-CoV-2 DNA available to labs on request. A few years earlier, DNA synthesis companies had banded together and created a select list of some sixty genetic agents, including those of the SARS and MERS viruses, that would require a security vetting of the customer before shipping.[16] The consensus by the time of the pandemic's second year, however, was that as the SARS-CoV-2 genome was already circulating freely, the benefits of making virus copies available outweighed the risks. But was the publishing of virus genomes a sound practice? A 2022 article in *ACS Synthetic Biology*, "Making Security Viral: Shifting Engineering Biology Culture and Publishing," offered three recommendations. First, journal editors should include security expert review before publishing viral study protocols. Second, that review process itself should be published. Third, article review should include a questionnaire asking authors to reveal their own security evaluations of their work.[17]

Other research and new product development progressed as several companies continued with research programs beyond pandemics. Using CRISPR engineering, the Cambridge, Massachusetts–based company Sherlock Biosciences had created inexpensive Zika and Ebola tests but swerved its entire workforce toward the coronavirus. By the end of 2021, it was shipping some 10 million COVID-19 tests a month to China, India, and Saudi Arabia, under an emergency use authorization.[18]

Such successes helped drive synthetic biology's public offerings. The stock market continued to rise, and the temptation was to raise money on the wave. In late 2020, Twist went public with a successful initial public offering (IPO) valuation of $110 a share, earning $300 million from the sale. In June 2021, Zymergen, with an IPO of $31 a share, raised $500 million. But such money could backfire. The danger was to overhype results. In fewer than eight months, Zymergen admitted in August 2021 that its revenue fell far short of predictions and would remain so. Its one product, a foldable phone screen material called Hyaline, was not performing well and had attracted fewer customers than anticipated. Its stock plummeted some 70 percent in a single day, and the CEO resigned. By 2022, the company was seeking a buyer.

As an industry, synthetic biology remained a realm of mostly small-to-moderately-sized companies and labs, or divisions within corporations, confined to small market shares while also performing research on behalf of government. The prospects were bracing, however. Some $3.3 billion was spent on tissue engineering research in 2021, for instance, but that figure was expected to leap to more than $26 billion by 2027.[19] Despite the mRNA vaccine success, the bid to expand and fully automate and become an industrial revolution remained a ways off.

As for the COVID-19 crisis, the synthetic biology company Distributed Bio, working with the World Health Organization and the U.S. military, was seeking to develop a universal vaccine platform through use of artificial intelligence, while continuing its search for new antibody treatments. In January 2021, Charles River Associates acquired the company, infusing much-needed cash. The synthetic biology company GenScript was offering its high-tech coronavirus test free to researchers. Hundreds of synthetic biology companies scrambled to meet the multiple demands of a pandemic in an uncertain future. Treatments, some created using synthetic biology technologies, were improving survival rates. Among these were antibody cocktails.

ANTIBODY COCKTAILS AND ANTIVIRAL PILLS

Synthetic biology contributed more than vaccines to easing the pandemic. Monoclonal antibodies were the laboratory-made molecules that mimic the immune system's defenses against a virus. In the case of SARS-CoV-2, the laboratory monoclonal antibodies targeted the virus's spike protein, either preventing it from binding to the patient's cells or labeling it to be destroyed.[20] To get such COVID-19 medicines, researchers isolated antibody-producing B cells from recovered patients' blood and reproduced the cells in humanized mice (mice engineered with functional human genes or cells). The Y-shaped antibody protein's two arms locked onto infected cells to prevent the infection from spreading. As the coronavirus mutated, however, it became necessary to devise multiple antibodies in cocktails to hit more than one target.

The initial COVID-19 antibodies were made by Regeneron, which received emergency FDA approval for use in patients with severe symptoms, such as President Trump when hospitalized at Walter Reed National Military Medical Center in October 2020. The way Regeneron used synthetic biology was as follows: First, the target was identified. Then, humans with resistance to the target were found through the Regeneron Genetics Center library. Medicines were then tested with a technology called VelociMab (mab for monoclonal antibody). The best candidates were tested in patients and the medicine optimized and produced in huge quantities in fermenters up to 10,000 liters in size.

In all, four such antibody treatments were approved for emergency use, with hard-to-remember names such as Bamlanivimab, etesevimab, casirivimab, and imdevimab. The government bought 1.5 million doses of Regeneron's medicine, but it did not help as much as hoped as the deadlier Delta strain of SARS-CoV-2 became widespread.[21] Cumbersome to administer, Bamlanivimab's FDA approval was revoked in April 2021.[22]

More significant were antiviral pills made by Pfizer (Paxlovid) and Merck (molnupiravir), with the hope of improving recovery as the highly communicable Omicron variant spread through the beginning of 2022. High-risk patients could obtain the pills with a prescription, but they had to be used early after infection to be effective. Learning from the widespread unhappiness with unequal vaccine access, both companies signed deals with the United Nations nonprofit Medicines Patent Pool to make the pills available at low prices in developing nations.

The Pfizer treatment initially required thirty pills over five days, and the Merck treatment required forty pills. Pfizer announced its pill was 89 percent effective if given within the first three days of symptoms, and Merck's pill was 50 percent effective.[23] The Pfizer pills were developed originally during the SARS epidemic, while Merck's were developed for flu. As early as 2015 Vanderbilt University researchers were testing Merck's molnupiravir for coronaviruses, where it seemed highly efficient as a treatment. Both were protease inhibitors of the class of drugs used to treat HIV, and Merck's was a polymerase inhibitor similar to hepatitis C treatments.[24] These kinds of inhibitors block a critical enzyme a virus uses to replicate itself and reduced AIDS death tolls significantly. The Merck pill, for instance, prevented the maturation of the virus and caused it to mutate. Pfizer's Paxlovid is a three-pill dose that consists of two separate medications. The first two pills inhibit an enzyme the COVID virus needs to make functional virus particles. The third pill is a booster.[25] The governments of the United States and United Kingdom both bought millions of doses of each, as it was hoped the pills would ease the worst of the pandemic. Swiss company Roche was working on similar antiviral medications.[26]

In the end, COVID-19 accelerated synthetic biology research.[27] As new variants like Omicron emerged, the biggest challenge was to distribute the biomedical breakthrough gains fairly, widely, and efficiently. Toward that end, discussions began for a new public health global network to make science breakthroughs more readily available to all.

A GLOBAL NETWORK

The effort that enabled Moderna and Pfizer researchers to create, test, and manufacture their mRNA vaccines in record time was a joint government, nonprofit, and business affair. When the Food and Drug Administration issued emergency use authorizations for the vaccines, it authorized that vaccinations would be covered by Medicare or be free of charge. Vaccinations in the United States began on December 14, 2020, barely eleven months after the virus was identified.[28] That vaccine triumph was only the beginning of what was needed.

Questions were raised about fairness. The manufacturers had worked with U.S. federal and state officials to get the medicines out "as quickly as possible," according to the Centers for Disease Control and Prevention (CDC) website. The rollout was complicated by the fact that storage required powerful freezers, and the vaccines had to be offered in two doses several weeks apart. Vaccine shipments were tracked online by the CDC's Vaccine Tracking System (VTrckS).[29] But access to these benefits was highly unequal. In the United States, if one was Latino or Latina, that person was twice as likely as non-Hispanic whites to die of coronavirus. If one was Black, three times as likely. Even after the vaccine rollout quelled the infection rate in Brazil and India, the pandemic raged nearly out of control in those countries.[30] A person's income played a key role in his or her chances of survival.[31]

As the pandemic surged again in early 2022, protests of the profits made by Moderna and Pfizer continued. The fortunes stood in contrast to past vaccine creators such as polio's Jonas Salk and Albert Sabin, who each earned nothing from their triumphal efforts. "It belongs to God," Salk famously said. In part, the windfall was partly a result of taking an enormous risk, and partly a result of larger economic changes since the Reagan era, when capital won out over labor, enabling people like Gates, Bezos, Zuckerberg, and Thiel to become multibillionaires. From an economic standpoint, the vaccines revived global markets at a time of world recession and would be considered worth the fortunes they generated. From a human standpoint, they offered the infinite value of saving a life.

But the question of profits loomed with each passing call for more vaccine availability, and thus sales. As BioNTech's market value soared past $21 billion, its founding couple, Ugur Sahin and Özlem Türeci, became among the richest people in Germany.

The boosters caused Moderna's stock prices to shoot up 35 percent in December 2021, leading Bernie Sanders to tweet: "This is obscene. Last week, 8 investors in Pfizer and Moderna became $10 billion richer as news about the Omicron variant spread. It's time for these pharmaceutical companies to share their vaccines with the world and start controlling their greed."[32]

Still, the carless couple Sahin and Türeci lived in a fairly unassuming Mainz apartment with their teenaged daughter, riding bicycles to work. "Discussing business is not his cup of tea," Pfizer's Albert Bourla told the *New York Times* of Sahin. Sahin announced he wanted to make sure the supply system could reach everyone, vowing to create "a global network" of regional manufacturing centers, beginning with one in Singapore and following with another in South Africa.[33]

As many fortunate people like me waited for their appointments, many of those in developing countries could still not get them or settled for the less effective Chinese and Russian vaccines.[34] Pretty much all I wanted was a vaccine, for family, friends, and me, but the fairness arguments were only beginning.

SUNSET ON THE BEACH

Inside the Bonita Beach Grande Crossing Publix Supermarket in southwest Florida, I could overhear an older woman talking about Chicago newspapers with another woman, possibly her caregiver. She was called up to the supermarket pharmacy counter. Next to go was a young man in gym clothes, baggy long pants and T-shirt, reading his cell phone. Then it was my turn.

"Theodore, what's your last name?" a young tech called out from the drive-up line. "Are you ready? You look so ready!"

The head pharmacist, Chad Slocum, worked nimbly with the components of the shot. "You were early, Theodore," he remarked.

It was April 13, 2021, in a strip mall supermarket beneath a sunset-rose-colored sky. I had pulled up with my wife to get my Moderna dose, parked, and walked into the pharmacy corner. It was not crowded. Two young women attended a makeshift reception folding desk set up near the cleaning products. "Let me see your identification," one said. "Not that" (I put away my Illinois driver's license), "your Florida ID." I produced the requested documents, following the state website guidelines.

After the shot—administered in the midst of handling drive-up pharmacy customers—I felt a flood of gratitude. Heading into the parking lot, I drove straight to the beach in the setting sun, thinking of enjoying the waiting glass of retsina wine with my wife and her high school friend at her friend's condominium. Hundreds of thousands like me were tweeting pictures of their vaccine shots, and I felt hopeful. Possibly it could not last.

"AS DISRUPTIVE AS IT GETS"

As more vaccines rolled out, the news carried reports about the fortunes the Moderna and Pfizer vaccine makers had made, along with the backers who had invested in them. In 2020, Moderna stock rose 7.2 percent in one day in December 2020.[35] The forty-seven-year-old Stéphane Bancel's 9 percent stake in the company had risen 50 percent in value since December 2018. This meant that, according to an estimate in *Forbes* magazine, he was now a billionaire.[36] So was MIT professor Robert Langer, who was a company board member and investor. Nobody suggested the profit was illegal. Rather, the question was whether such profits were proper or ethical in a time of global depression.

The U.S. government invested in Moderna, which then sold its taxpayer-funded vaccine back to the government. Although Pfizer did not take U.S. federal money at first (only later, to help meet demand), issues were raised about the costly packaging of the vaccine. The larger forty-eight-well containers used required well-equipped industrial facilities generally more available in urban sites, enabling Pfizer to control the distribution to profitable markets, it was suggested.[37]

BioNTech's well-liked founders, Özlem Türeci and Ugur Sahin, earned a similar windfall.[38] They were multimillionaires already, following the sale of their previous 2001 biotech startup Ganymed, which sold for $1.4 billion in 2016. They had taken their successor BioNTech public early, at a seemingly inopportune moment, in October 2019, right before the coronavirus outbreak. Its initial IPO was a disappointment.[39] but by June 2020, the stock price had tripled in value and continued to rise through the fall. Other companies gyrated in the opposite direction. Zymergen's August 2020 crash was followed by an investor lawsuit. On the other hand, Ginkgo Bioworks' Wall Street offering raised some $1.6 billion, and the company started two new foundries in Cambridge to augment the Boston Design Center space it had outgrown. Overnight, the five cofounders became multimillionaires, close to billionaires. "Jealous," tweeted Twist Bioscience CEO Emily Leproust, as synthetic biology followed its biggest investment year in 2020 with a bigger one in 2021 and a larger one in 2022.[40] In July 2022, Ginkgo bought Zymergen.[41]

The pandemic was profitable to the vaccine makers, to be sure. With sales of $21 billion for Pfizer's vaccine alone, the coronavirus vaccine became the biggest-selling medicine in history. Both Pfizer and Moderna prepared to extend the success of the messenger RNA platforms to treat other infectious diseases, such as flu, tuberculosis, rabies, malaria, Zika, and viruses that had not even emerged yet. Each company had an mRNA cancer program in place as the Biden administration proposed a massive investment in biotechnology "This is about as disruptive as it gets," the University of British Columbia's Pieter Cullis told the *Washington Post*.[42] "It's a fantastic time for life science," added Cullis, who helped create the lipid nanoparticle casings for the vaccines, a technical innovation some ranked as high as that of the mRNA vaccine itself.[43] But how do you share the benefits of taxpayer-funded research that yields such high value?

As new SARS-CoV-2 variants spread, it seemed like masks, lockdowns, and distancing might become recurring fixtures of our world. When focus on the new science increased, the ethics issues had to be confronted. The Biden administration bought hundreds of millions more doses of the Moderna and Pfizer vaccines and planned to distribute them to poor and developing nations. In 2022 BioNTech unveiled a mobile container-sized vaccine factory that could manufacture doses in Africa within the year.[44] But that was not enough. The benefits had to be shared, domestically and internationally.

Synthetic biology came of age with the global production of vaccines. Many, however, questioned the higher prices of meatless meat, fishless seafood, and milkless dairy products. To shape the public debate, the field's leaders, and their communications officers, became major figures on Twitter, podcasts, Zoom casts, and in publications from *Fast Company* to *Forbes*. But some of their arguments changed.

In the wake of the pandemic, some synthetic biologists no longer claimed they were simply industrializing biology. Humans had invented other industries. But biology had invented humans.[45] Biology was not a mere technology, but rather a majestic sphere beyond human activity, almost of creation one might say. In enlisting biology's help, companies were the partners of a profound collaboration, in which universities were no longer the centers of technical innovation. "Companies are incredible vehicles," Ginkgo's Jason Kelly told investors in a January 12, 2022, Zoom conference, "to tackle big things."

The speed of such advances made it time to examine the ethical and policy debates surrounding the new science, including viral research, CRISPR gene editing of human embryos, and synthetic biology's unequal distribution of benefits in a world of rich and poor. As lab techniques became widely available, fresh questions arose about biosafety and bioterror. How would DNA synthesis companies know who was ordering genetic material, and for what purpose? Garage science rebels joined an informal, do-it-yourself secret society of biology hackers. The benefits accrued from taxpayer funding were unequally distributed, they argued. It was time to understand the responsibility of taxpayer-supported science research to the public need.

13

THE MOIRAI'S GIFT

The last speaker at the conference is always the ethicist.
—Laurie Zoloth

A group of researchers had gathered by Zoom to consider the possibility of opening synthetic biology to
indigenous forms of knowledge. The group, including
the Faber Futures biodesign company director Natsai Audrey
Chieza from London, discussed the benefits of tapping many
more diverse voices outside of the classroom and boardroom in
charting a course for science research.

Around the world, the pandemic had revealed the dangers
of unequal access to health care. Protests had mostly stopped,
but many people were worried about inflation and disenchanted.
Issues of access to the COVID-19 treatment and prevention breakthroughs, and the to the benefits of biotechnology,
remained. As the SARS-CoV-2 virus was mutating and vaccine boosters were being administered, many in the developing
world could not get shots. The vaccine companies refused at first
to share their intellectual property with countries in Africa and
South Asia.[1] Some synthetic biology founders were becoming

multimillionaires, even billionaires, while much of the world waited for medicine. Some anti-vaxxers were taking expensive experimental treatments while emergency rooms in the developed world remained full of unvaccinated, and vaccinated, COVID-19 patients.

There were lessons learned from the way in which government-supported vaccine research was turned to astronomical profits, observed economics professor Massimo Florio of the University of Milan. "Governments should be stronger than corporate lobbying and should require both a patent waiver and deep technology transfer to qualified third-party producers," Florio wrote in a *Research Europe* opinion column. "We must define a modern form of intellectual property rights in the public interest."[2]

The lack of access to expensive biomedical breakthroughs, and the dropping cost of lab equipment, led to a do-it-yourself community proclaiming molecular genetics should be for everyone. Molecular biologists with PhDs and amateur activists were shooting YouTube videos on how to make insulin drugs or do personal gene-editing in garages and apartments. Some ordered kits on Amazon, while serious, idealistic scientists founded community labs dubbed "biospaces" in cities around the world. Some of them, such as New York–based Genspace, sponsored classes and built labs almost as advanced as those "generally confined to well-funded academic institutions and private corporations," Margaret Talbot wrote in the *New Yorker*.[3] Others made the technologies "better, cheaper, faster, and more available," said Opentrons' Will Canine, such that first-rate academic and commercial labs purchased them.

In the beginning, many in synthetic biology had sought universal accessibility. The BioBricks and OpenWetWare websites featured a free online library of standard genetic material.

The malaria medicine created in Jay Keasling's lab was made available at cost in Africa.[4] Those altruistic researchers of the 2010s sought solutions to the urgent problems of a new social and environmental order, ranging from climate change to unequal health care. By 2022, some of the biggest synthetic biology companies had patented discoveries and reaped increasing windfalls. The biggest exemplars of the windfalls were Moderna and Pfizer, who would not divulge their COVID-19 vaccine formulas, arguing that patents protected their hard work and imagination.

The argument over the origin of SARS-CoV-2 also shed new light on the risks of under-regulated research. A World Health Organization panel in 2021 called for stringent global review of access to gene editing data.[5] In the United States, the National Biodefense Science Board, planning for the next pandemic, climate change, and bioterror, called for a regulating body to keep pace with scientific advances. Synthetic biology was now a subject of wider media coverage. With a valuation of the biotechnology industry at more than $4 trillion, some companies promised more sustainable and affordable clothing manufacturing, including the one founded by Natsai Audrey Chieza and others like it. Still, the problems of access persisted.

In Greek mythology, the three Moirai (or Fates) write our stories, determining our life span by spinning and cutting the thread of life. They are remorseless. As the pandemic ebbed and surged, poorer regions of the globe struggled. How could science pay back the governments that funded it and the people who funded the governments? What was the best approach to bringing in new voices and ideas to science? It was time to answer those questions.

"THE MOST STUNNING ETHICAL EVENT IN THE HISTORY OF SCIENCE"

In 1974, a federal research biochemist we have met briefly, Maxine Singer, was worrying about the misuse of the powers of genetic research. She chaired a conference where the latest techniques of recombinant DNA got her thinking about the danger posed should gene-edited pathogens leak from a lab. She and her conference co-organizer wrote a letter to the National Institutes of Health (NIH) expressing their concern that "too little solid information" existed to allow such DNA research to proceed without monitoring.[6] At the suggestion of the NIH, the Conference on Recombinant DNA Molecules (commonly referred to as the Asilomar Conference) met in 1975 at the Asilomar Conference Grounds in Pacific Grove, California, to consider a way forward.[7]

Lurking behind the calls for ethics review was a sometimes-horrific past of medical research abuse. While the ethicists could draw upon a tradition of thought from Hammurabi and the Talmud to Aristotle and Hippocrates, Muhammad and Maimonides to Susan Sontag, they also had to confront a historical record marred by terrible exploitation of human subjects. That history included Nazi experiments on concentration camp prisoners; the United States Public Health Service and Tuskegee Institute's mistreatment of Black inmates, some of whom were told they were receiving antibiotics for syphilis when they were not; forced sterilization of some Latina women; and numerous transgressions when the disenfranchised were subjected to research without consent.[8] Such breaches led many universities to create institutional review boards in the 1980s and 1990s to assess experiments involving human beings.

For three days at Asilomar, arguments raged between prominent figures, including the three future Nobel Prize winners Sydney Brenner, Paul Berg, and David Baltimore (the latter also a future National Institutes of Health vaccine committee director), as well as Maxine Singer, who would go on to become president of the Carnegie Institution in Washington, D.C. Ultimately participants agreed that research should continue "under stringent restrictions," Berg recalled.[9] Their proposal confined the gene modification of pathogens to the laboratory and prohibited the injection of pathogens into human beings and the editing of embryonic genes. Other rules focused on the protection of privacy, proper transparency about risk, and an inclusive and ethical system of choosing and communicating with human subjects for clinical trials and medical research. These guidelines became a basis for a U.S. federal guide for publicly funded research. Berg, introduced in chapter 5 as the first researcher to transfer a gene from one organism to another, worried most about science privatization. "Once . . . corporations begin to dominate the research enterprise, it will simply be too late," he warned colleagues.[10] The NIH created a standing Recombinant DNA Advisory Committee to address that concern.[11]

Amidst this oversight, organism engineering advanced so rapidly that Hollywood thrillers like *Blade Runner* or *Gattaca* became the sounding boards for social anxieties. New fields of scholarship, including disability studies, medical humanities, and animal rights theory, arose to explore how our words shaped our willingness to harm animals or each other in the guise of science. At the same time, the infusion of capital into biomedicine continued. "The commercialization of molecular biology is the most stunning ethical event in the history of science," cautions the narrator of the 1990 novel *Jurassic Park*. "Suddenly it seemed as if everyone wanted to become rich."[12]

Some of the people working in synthetic biology companies fell in love with *Jurassic Park* when they were children. Today, Jurassic dinosaur stencils and statues decorate the automated foundries of synthetic biology company Ginkgo Bioworks. Steven Spielberg gave a dinosaur head to Ginkgo's CEO when the company went public. Thus two threads, one academic and the other popular, triggered awareness of the need for ethics review of synthetic biology research.

A CHRONICLE OF SYNTHETIC BIOLOGY ETHICS

From the start, many synthetic biology researchers deemed ethics a cornerstone of a democratic science guided by a respect for life. Many of the field's founders, like Stanford's Drew Endy, MIT's Tom Knight, and Harvard's Pamela Silver, argued that findings and technologies should be open source. To some newcomers, Endy became "my leading spiritual, ethical guide on what we should build," said robot maker and Opentrons CEO Will Canine. In the same idealistic vein, the student iGEM competition required all experiments to contain a human impact statement. "I have never seen a science so focused on ethics," Target Malaria researcher Alekos Simoni, coauthor of an article proposing ethical guidelines for gene drive studies, told me.

As investment in synthetic biology increased, however, such idealism clashed with the demands of stock pitches. Universities lost some of their frontline research positions to commercial laboratories where there was little ability to monitor or police research practices.

The speed and range of researchers' power to design and manufacture new life forms continued to increase. The J. Craig Venter

Institute's first semisynthetic bacterium prompted the Obama administration in 2010 to ask the Presidential Commission on the Study of Bioethical Issues to create a report on the ethics of engineering life.[13] The commission declared that synthetic biology research should be based on the principles of public benefit, responsible stewardship, democratic discussion, and fairness of availability. This report was followed in 2012 by a federal National Bioeconomy Blueprint to serve as a research road map.[14] But research kept speeding forward. The advent of CRISPR gene editing in 2012 was followed by rapid improvements in DNA synthesis, as hundreds of labs fine-tuned genes inexpensively and quickly to make medicines, clothing, food, new organisms, or living computers. Babies born with edited genes followed in 2019. Such breakthroughs brought to life the pages of a science-fiction novel as policy makers struggled to keep up.

One set of responses came from the community, where various do-it-yourself movements each gained its own following.

MAKER SPACES

In a red brick building near Gowanus Bay, the Brooklyn-based community lab Genspace is reopening for classes, "where anybody, anywhere can do science," offers executive director Beth Tuck. The oldest community biology lab seeks to demystify science and allow community members to learn such techniques as sequencing their own DNA or designing their own organisms. As a part of the movement to bring easier access to lab technologies such as CRISPR gene editing, community laboratories like Genspace have spread around the world—offering conferences, classes, and videos encouraging people to try out cutting-edge lab work themselves.

It began with a group of entrepreneurs, artists, and scientists who met in a New York apartment living room in 2009. They wanted to let the general public learn science. Building on the open science advocacy of the 2000s and synthetic biology's annual iGEM student competitions, which featured a category inviting community lab teams to participate, like-minded thinkers sought to democratize science by opening access to technology such as polymerase chain reaction (PCR) machines to clone genes. Other community labs soon followed, such as La Paillasse in Paris, France, MadLab in Manchester, England, Bioligigaragen in Copenhagen, Denmark, and BioCurious in Santa Clara, California. By 2020, there were more than sixty such community labs, with the number increasing every year. At Genspace, the goals are "ethics, transparency and diversity," achieved by classes and tutorials sparking a sense of "curiosity, experimentation, and collaboration."[15]

One part of the motivation could be dated to 2015 when the nonprofit collective called Open Insulin in Oakland, California, protested the high price of insulin in the United States, produced by the companies Eli Lilly, Novo Nordisk, and Sanofi, as it rose from $21 to $300 a vial in a few years. There was no economic reason for the increase, argued Colorado State University biochemist Jean Peccoud, founder of the journal *Synthetic Biology*, "no justification other than greed."[16] Buying up expensive equipment on eBay from bankrupt biotech companies, by 2018 the biohackers were featured in multiple news stories. Leading spokesmen included the former NASA biochemist Josiah Zayner, who injected himself with CRISPR-edited genes on YouTube to increase muscle mass. "My mom," he admitted to *The Guardian*, "was so sorry I left NASA."[17]

Some of the directors were professors. Genspace was developed by Ellen Jorgensen in Brooklyn, New York, formerly an adjunct professor in the Pathology Department of New York Medical College. She taught Genspace's free course and also founded the nonprofit Biotech Without Borders. It was she who incubated the Genspace's first genetically edited fluorescent *E. coli* under her armpit while in bed at night. From MIT's Media Lab, David Sun-Kong directed the Cambridge, Massachusetts–based Community Biotech Initiative, to make the new science available to all. Sun-Kong was also an award-winning rapper.

The robot-making company Opentrons modeled a partnership among community biologists, public health officials, and venture capitalists. It too began at Genspace. Cofounded in 2013 by political science major Will Canine and New Jersey robot-maker Chiu Chau, the company produced an easy-to-use, high-quality liquid-handling robot to make DNA analysis more affordable. Located in a brownstone alongside artists, where the doorbell sounded bird calls, the company won a start-up grant from Y Combinator, the seed investor that had supported companies such as Ginkgo Bioworks and Dropbox. It participated in the iGEM student competition and "received orders from the top synthetic biology labs around the country," Canine told me.[18] Opentrons expanded manufacturing facilities in China and the United States. After the company won a competitive bidding call by New York City to test people for COVID-19, hospitals around the world snatched up its robots. The company's valuation rose to $1.8 billion. "There are some magical opportunities where a venture-backed business and a good political project can overlap," Canine explained. "Opentrons is one of those."

The fact that wealthy nations needed do-it-yourself (DIY) biology, however, was criticized. DIY drug manufacturing in a nation like the United States signaled that we had "a broken system," Yale epidemiology professor Gregg Gonsalves told the

New Yorker. It showed the "dead end of desperation," of people who could not afford high-tech health care. "All that energy and anger might be better focused on politicians," Gonsalves continued, to improve access to the latest care.[19]

Toward that end, new policy proposals sprang up to confront unequal access to medicines. A new academic field, called health humanities, sought to fill the gaps when high-tech Western medicine failed in reaching the people who needed it most. A new type of physician, the hospitalist, was designated to coordinate patient care among different specialties and the unequal power relation between doctors and the people they served.

One of the first tasks of the medical humanities, and of ethicists, was to give more scrutiny to gene editing of humans.

GENE EDITING REVIEW

As the two Chinese girls with edited genes turned three years old, the ethics committees of several science organizations in 2021 promoted new policies of gene editing oversight.[20] The World Health Organization was one. A similar set of guidelines was developed by a combined group at the U.S. National Academy of Sciences, the U.S. National Academy of Medicine, and the Royal Society. Both proposed that researchers slow down or stop the editing of human embryos, which cannot consent. The Russian geneticist Denis Rebrikov, researching gene edits for children of deaf parents, committed to following the new guidelines, but cautioned: "Where did you see the researcher willing to slow down?"[21]

The popular tide was turning for some gene therapy in humans, where safer and more precise techniques such as base editing, altering only one base (A, C, T, or G) of a genome might relieve suffering. Several biotechnology companies were working on gene therapies for congenital deafness, and thousands of people with sickle cell anemia were anxious to join any gene editing study they could find.[22] Updated rules for gene therapy studies were discussed in March 2023 at the Third International Summit on Human Genome Editing in London, but a final report was not completed.[23]

One model could be found in the regulation of stem cell research. Today, pluripotent stem cell researchers, like those at the company Novo Nordisk, may only use human embryonic stem cells "derived from surplus embryos from in vitro fertilization (IVF) treatment that are donated with freely given informed consent." Such embryos are otherwise "destined by law to be discarded," according to the company website.[24] Even so, stem cell regulations spawned from President George W. Bush's Council on Bioethics were strict.

Back in 2012, scientists had issued a report titled *Principles for the Oversight of Synthetic Biology*, which included the banning of human germline editing and protection of workers and the community. It mentioned something called the "precautionary principle" invoked since 2001 in environmental research. The principle included four elements, according to an *Environmental Health Perspectives* article written by Massachusetts researchers: "Taking preventive action in the face of uncertainty; shifting the burden of proof to the proponents of an activity; exploring a wide range of alternatives to possibly harmful actions; and increasing public participation in decision making."[25] Another review article in 2021 suggested many of the same ethical guidelines for gene drive research.[26]

With such guidelines in formation, an immediate challenge was to monitor the lab virus study we reviewed in chapter 11, "gain of function."

NEW VIROLOGY OVERSIGHT

A joint international effort was under way to create comprehensive limits to gain-of-function study of viruses. Giving viruses additional functions had, in many cases, produced successes. In 1937, researchers studying yellow fever devised a vaccine by causing the virus to infect chicken cells. A herpesvirus had been engineered to attack human cancer cells and was now an FDA-approved therapy for melanoma.[27] Many gene therapies are delivered by engineered, weakened viruses. But if there was one thing that all could agree on after the pandemic, it was that any future gain-of-function virus experiments must be better regulated.

As far back as 2006, the National Biodefense Science Board (NBSB) was created by Congress, with thirteen members from academia, government, and business, to protect the insights gained from the study of pathogens. In 2016, the U.S. National Science Advisory Board called for an international governing body over molecular biology to ensure that gain-of-function research was to be strictly overseen and openly conducted.[28] The criticism of gain-of-function study of respiratory viruses stems from a University of North Carolina study by Ralph Baric in 2015, when his team took the first SARS virus, added a surface protein from a horseshoe bat virus, and showed it could infect mice.[29] In January 2020, the NBSB created an expert panel to review rules for such gain-of-function virus research.[30] By 2022, a National Security Commission on Emerging Biotechnology was being formed to address new threats and, by January 2023, stricter guidelines on gain-of-function research were proposed to the National Science Advisory Board for Biosecurity (NSABB).

Several companies, such as Twist Bioscience, instituted internal institutional review boards and formed red teams to try to hack their own biosecurity measures. The Boulder-based digital gene company Inscripta ran algorithms to determine whether customer DNA requests were safe. "We need better tools for understanding function from gene sequence," Inscripta's biosecurity specialist Elizabeth Vitalis told me, "when we create life forms that have never been seen before."[31] Experts agreed that an international registry of dangerous gene sequences accessible only to qualified researchers was needed, but how best to maintain the secrecy of such a registry remained undetermined.

SCIENCE FOR EVERYBODY

As 2022 rolled on, Ginkgo Bioworks was valued at an astounding $17.5 billion" with: "was planning a new 228,000 foot facility called the Foundry at Drydock. In a May 11, 2021, filing to go public as a special purpose acquisition company (SPAC), Ginkgo had called biology "the most powerful manufacturing technology on the planet." The SPAC combined Ginkgo with the Soaring Eagle Acquisition Corporation, providing the new company with $2.5 billion in cash. The new company was named Ginkgo Bioworks Holdings. Jason Kelly and Reshma Shetty remained on the board, where they were joined by experienced public company investors like Harry Sloan, former CEO of Metro-Goldwyn-Mayer (MGM).

A special purpose acquisition company is a publicly traded company with a two-year life span formed to effect a merger enabling another company to go public. SPACs raise money mainly from public-equity investors and often offer a company better terms than a traditional IPO would. SPACs had a popular run. In 2020, 247 were created with $80 billion invested.[32] In the first quarter of 2021, 295 were created, raising $96 billion.

They are generally formed by investors with expertise in one field. Ginkgo was a unique opportunity in the founding of a new industry, argued Sloan and Kelly, in a "category of one (that) launched the modern practice of synthetic biology."[33] The company made money by user fees and by taking partial ownership in its clients, such as Motif Foodworks, Synlogic Therapeutics, Huue (a sustainable clothing dye manufacturer), or Arcaea (a sustainable cosmetics maker). Ginkgo's 2021 IPO raised $1.5 billion.

To address issues of equity, Ginkgo began offering new employees class B shares, making them part owners and giving them greater voting power than the investors in the original class A shares. The company published a magazine, *Grow*, edited by Christina Agapakis, to explain "the unfolding story of synthetic biology" to the public. *Grow* devoted its entire October 2021 issue to equity in the field, featuring long-form articles on unequal access to COVID-19 vaccines north and south of the border in Laredo, Texas, and a history of the 1970s Black Panthers public health advocacy in Black communities, to name a few of its thoughtful treatments of hidden science history.[34]

From a lawn chair on the grass, my DePaul University colleague, health humanities professor Craig Klugman, discusses the ethics of treating people who refused to be vaccinated. Such people have to be treated, he tells my freshman class. It is a rule of Hippocrates. Then he explains why we are sitting outside in the September Chicago sun, wearing masks, and not in the classroom. A member of the National Biodefense Science Board, Klugman works on measures to prevent the next pandemic.

COVID-19 confirmed for the world the power of biology, and of synthetic biology. The pandemic brought questions of access to science into sharp relief, as our world of expanding human contact with animals and climate change met up with mutating viral pathogens. Suddenly a bigger audience understood the

reach of such viruses, their antibodies, and vaccines. A ratio-
nal observer would have been gratified that government could
promote research and make its benefits, like vaccines, available
free of charge. Tasked with preventing more crises, the National
Biodefense Science Board was meeting quarterly under Xavier
Becerra, secretary of the Department of Health and Human
Services. Together with the secretary of state, Antony Blinken,
Becerra wrote an opinion piece in the *Journal of the American
Medical Association* calling for strengthening the World Health
Organization's pandemic response powers, the international
sharing of data, and increased funding to developing nations to
help detect and contain outbreaks earlier in their spread.[35]

Leading thinkers decried the high cost of genetic cures for
illnesses such as Huntington's disease (about a million dollars for
a single surgery) or sickle cell anemia. Ethicists shunned the idea
of editing embryos of the unborn and devised rules for one-time
correction of gene defects. Dalhousie University's Francoise
Baylis raised concerns about gene editing, gaining the support
of such disparate voices as Harvard's technological cheerleader,
biologist George Church, and the Divas with Disabilities Project, science critics who worried about the ability of technology
to direct the fates of "those most vulnerable for extinction."[36]

Laurie Zoloth, a University of Chicago professor of religion
and ethics, connected Jewish themes from the Talmud with
arguments on behalf of science democracy, diversity, equity, and
mutual appreciation, along with a plea for community involve-
ment in research decision-making beforehand, rather than regulations or punishments afterward. She published an influential
2021 *Cell* article, "The Ethical Scientist in a Time of Uncertainty," entreating researchers to "cultivate classic values of verac-
ity, courage, humility, and fidelity."[37] The gene editing dividing
line, most ethicists agreed, was to permit somatic or body cell

gene editing for curing single-gene diseases like Huntington's or sickle cell anemia, but stop at embryonic editing.

A few took a more extreme position in favor of embryonic gene editing, including Julian Savulescu, of Oxford University's Uehiro Centre for Practical Ethics. Savulescu argued that gene editing is a tool of nature that should be made available to parents at risk of passing on genetic disease to their children. Not to use CRISPR gene editing would be bad parenting, in Savulescu's view.

Today, a new concern is a currently permissible advancement that enables parents using in vitro fertilization to choose from multiple embryos the one with the most desirable genes. Elsewhere, in 2021 the U.S. National Counterintelligence and Security Center issued a report warning that a prenatal genetic test from the Chinese company BGI was being used to collect data from millions of unborn babies.[38] Gene harvesting and theft are practices by which unscrupulous researchers may tap the richness of data from humans and the biodiversity of undeveloped nations without informed consent. One related question is whether future parents will be more tolerant of embryonic gene editing than those in the past. The ethics of the age of biotechnology presents great opportunity and some significant risk.

BRIGHT LIGHTS

Setting out from my daughter and son-in-law's San Antonio home, I push my granddaughter's carriage past the river haunted by ghosts of the Spanish. Their seventeen-month-old sits wide-eyed in her stroller. It is hot. As a new grandparent, I feel that Earth's future is my number one priority.

This was synthetic biology's moment to change that future. Some companies were pushing for a circular economy. Some sought to turn waste into energy. Some were toolmakers. Some sold DNA. Meat without animals, seafood without fish, milk without cows, eggs without chickens, and microbial sensors to assist your immune system, joined in a race for solutions to the world's most pressing problems.

By the fall of 2022, the World Health Organization called for an international registry and standards for all human gene editing experiments, as well as an avenue for whistleblowers to speak out. The report opposed unscrupulous sales of faulty treatments to those in chronic pain and the use of gene editing to enhance athletic or intellectual abilities. Several gene editing trials were under way related to cardiovascular and other diseases. Even then, many ethicists insisted on a mechanism for spreading the benefits of gene editing equitably to prevent the advantages going primarily to elite groups. Still, there was no method of enforcement, other than governments' own funding oversight.

Amidst the pandemic, observers pointed out the inequity in COVID-19 survival rates. In the Coronapod podcast of the journal *Nature*, journalist Amy Maxmen showed how vulnerable the rural poor were to phony cures and misinformation. The vaccines were not making enough headway into rural populations, such as the migrant farmworkers of California's San Joaquin Valley.[39] Some people mistrusted doctors and hospitals, feared the expense of treatments, and suffered alone, in some cases spending more for quack cures while ignoring a free vaccine.

The new industry stood poised to respond to the next global pandemic, with prospects of RNA treatments for diseases ranging from cancer to AIDS. One positive sign was an explicit acknowledgment of the potential of synthetic biology in the Biden administration's proposed science and infrastructure budgets.[40]

The CDC maintained surveillance of inbound pathogens at ports of entry, employing the Ginkgo Bioworks subsidiary called Concentric as one of its partners to do so.[41] But the spread of monkeypox, and the conflicting public health messages about it, had some critics saying the CDC had not learned enough from its COVID-19 mistakes.

Finally, under pressure from activists, the Biden administration announced it would partner with industry to spend billions of dollars to manufacture vaccines to reach poorer countries.[42] It was hoped that everyone who wanted a vaccine could get one, and that anyone who wanted training in the new industry could get it. The 2022 White House Executive Order on Advancing Biotechnology called for more educational programs in biotechnology at Historically Black Colleges and Universities, other minority-serving institutions, and in community colleges and technical high schools.[43]

Accessibility to biotechnology poses a global challenge. But first, a glimpse of the potential of programmable life comes from the vantage point of space.

14

TO THE PLANETS AND BEYOND

Synthetic Biology in Space

We can do darn near anything.
—Lynn Rothschild, NASA

At an astrobiology conference on a remote Sicilian mountain, University of Minnesota researcher Kate Adamala loved standing outside at night, watching the stars explode in the sky.[1] She had grown up reading Isaac Asimov and Ursula Le Guin books in her parents' small, Soviet-style apartment in Poland. Now she was working on life's origin in a leading Italian lab. She had never thought that was something you could do as a career, and get paid for it.

Many researchers saw space as a significant testing ground for miniaturizing and lowering costs of synthetic biology, which could then be applied to Earth environments. In space, modified yeast and bacteria could recycle waste into energy, fertilize soil to grow food plants, clean dirty water, and turn carbon dioxide into medicines. Some synthetic biology researchers turned to creating hardy forms of life to support colonies on Mars, starting with microbes from frigid Earth deserts. With plans to return to the moon and Mars, NASA and the European Space Agency reached further. German researchers modified bacteria to produce protein from dim sunlight or to seed Mars soil, and U.S. researchers sought to build storage and housing structures from freeze-dried mushroom roots.

The application of synthetic biology to space was spearheaded at the NASA Ames Research Center near Mountain View, California. At the smaller Center for the Utilization of Biological Engineering in Space (CUBES) in nearby Berkeley, other researchers from around the country were programming DNA to make ingredients for medicines, food, and materials in Mars-like conditions. A newer site of research was the European Space Agency's Barcelona-based Micro-Ecological Life Support System Alternative (MELiSSA) pilot plant which produced, among other things, radiation-resistant plants that could be grown in tiny spaces with minimal sunlight and water. Elsewhere, as on the top of Hawaii's Mauna Loa volcano, NASA was studying the effects of an experiment in which six young people were isolated for eight months at a remote site, communicating with "Earth" only on a twenty-minute delay, as if on Mars. "I miss it," their captain, James Bevington, said to me of the peace he felt there. "I loved it."[2]

From food and energy creation to repair of building materials, space scientists engineered microbes and plants that could help sustain life on other planets. "Synthetic biology is core to NASA's mission," MIT's Chris Voigt told me. Some leapt at the challenges as an opportunity for fundamental transformation. One significant name among those inspired by synthetic biology in space was that of NASA's Lynn Rothschild, lead researcher of NASA's Bio and Biology-Inspired Technologies.[3] Another was Kate Adamala, cofounder of an innovative research company and leader of an international group to create synthetic cells from scratch. "Life's not a hard technology," Adamala told me.

"It's made of proteins, lipids, and other molecules, and it is programmable."[4]

The venue was space, but the small, lightweight, living technology could be applied at home. "The social good would be to use these technologies to feed, house, and clothe people on Earth," said Adam Arkin, Berkeley biologist and CUBES director, "using just carbon dioxide, light and water."[5] Understanding synthetic biology in space could help us to save Earth. To understand how, we begin in the imagination and then journey to Mars, a Hawaiian volcano, and a snowy Colorado mountaintop.

IMAGINED WORLDS

For 150 years, science-fiction writers have imagined spaceships loaded with technologies to make planets inhabitable for humans. That process, called terraforming, entails modifying a planet's natural state as humans have done on Earth. The term came from a 1942 short story in the magazine *Astounding Science Fiction*, Jack Williamson's "Collision Orbit," but the idea had been around much longer.[6] H. G. Wells's *War of the Worlds*, for instance, featured a reversal in which invading Martians try to terraform Earth to help themselves survive.[7] Author Robert Heinlein explored an agricultural alternative in his 1950s novel *Farmer in the Sky*, about terraforming Jupiter's moon Ganymede using the ringed planet as a sun. Early scientific ideas included the proposal that greenhouse gases could melt Mars ice into water. In 1973, Carl Sagan suggested doing much the same by transporting darkened material to the red planet's ice caps to absorb sunlight. In Arthur C. Clarke's *2001: A Space Odyssey*, aliens turn Jupiter's moon Europa into an escape haven, while casting down their black monolith for the seemingly promising primates of our planet.[8]

What many of these tales shared was hubris. Other novels raised the obvious ethical question: What right do we have to alter other planets? That question is familiar to *Star Trek* fans, where the prime directive is never to interfere with a planet's life forms. Beyond the doctrine of non-interference is the danger of transforming a planet into something resembling Earth's industrial wastelands. NASA researcher Chris McKay answered such questions in a series of scholarly articles and in a 2021 presentation at the Mars Society virtual conference.[9] McKay concluded that a planet might be terraformed ethically, but only if no native life forms were found first. If such life forms were discovered, then they are to be encouraged to thrive. Others argued for "para-terraforming"—altering a lifeless planet's conditions only in enclosures governed by international treaty.[10] Yes, but how?

Probably with a lot of synthetic biology, whether in covered plots shielded from hostile temperatures and toxic atmospheres or in underground quarters safe from solar radiation. How would you engineer life to grow in such environments to make medicines, materials, and food? You could modify existing life forms or build new ones of your own. That's a part of what one international group of researchers, called Build-a-Cell, have been discussing for years in weekly sessions chaired by Kate Adamala.

TO BUILD A CELL

Ever since she could remember, Adamala had been fascinated by extraterrestrial life. Growing up in a working-class neighborhood in a Polish city, she played with chemistry sets that sometimes exploded in the family's apartment, driving her mother crazy.

But working-class life had its advantages. Adamala's education was free. She loved thinking about how life can emerge from chaos and made it a focus of her studies.

Joining the team of biochemist Pier Luigi Luisi at Roma Tre University, Adamala traveled to conferences he chaired in India for discussions with Buddhist monks. The professor booked their team sometimes in a tiny monastery, at other times in a posh mountain resort. "Why does DNA always try to replicate itself?" the monks might ask. Good question. From Luisi, she learned how to think deeply in multiple disciplines, starting in chemistry and then turning to biology because it "made chemistry come alive," she explained to me.

Adamala went on to Harvard where she worked with geneticist Jack Szostak on some of those big questions and others. She and Szostak published papers on the ways cells might form from organic materials. "Oh yeah," Adamala recalled in her droll manner, "and he won a Nobel Prize."[11] When Adamala's husband got a job in Boston, she had to find a postdoctoral position quickly. She e-mailed the bearded, outspoken neurobiology researcher Ed Boyden at MIT, and he agreed to hire her. With him, Adamala pursued cutting-edge work programming RNA to help heal damaged brains. She and Boyden cofounded a company called Synlife to modify cells to treat victims of combat injuries, car accidents, or the brain trauma of concussions, research they covered in a paper published in 2020 in *Cell*.[12]

From health to waste treatment, food to energy, engineered microbes could help sustain life in space. But here on Earth, the research field needed to unite multiple disciplines. To do that, Adamala joined with Stanford's Drew Endy and others to create an international group sharing lab insights. "The technologies had gotten good enough for researchers to unify," she told me, "toward building a living cell. That's amazing because synthetic

biology is not known for unifying!" The news media was skepti-
cal. The online science site STATNEWS called their Build-a-
Cell group "a motley crew of undergrads and legendary genetic
engineers."[13] Seminar topics included the Max Planck Society's
efforts to craft cell membranes in Germany, the University of
Bari Aldo Moro's program to create photosynthetic cells in Italy,
and the success of New York's Binghamton University in engineering bacteria to generate electricity. "Build-a-Cell blows my
mind with the amazing work going on in labs around the world,"
observed Lynn Rothschild, herself a leader in adapting synthetic
biology to space.

As part of NASA's own research support, the agency took
Build-a-Cell one step further, creating a university center for the
study of synthetic biology in space, which is where we turn next.

GRAVITY

You are on a planetary mission's sixth month, firing across space
in your cramped aluminum and graphite vessel. Bored and sick
of frozen burgers and protein shakes, you long for a taste of real
food. You cannot order from Grubhub. What do you do? Grow
your own, with synthetic biology. Luckily, NASA's Space Synthetic Biology project had developed several grow-your-own
food technologies for long space flights. One was the BioNutrient system, containing dehydrated yeast modified to make
essential nutrients found in vegetables, and which the agency
was testing on the International Space Station.[14] Just add water,
mix, keep the packet warm for forty-eight hours, and enjoy a
comforting, nutritious meal. The astronauts were conducting
multiple rounds of tests on the system, adding sterile water and
powdered food for the yeast. The BioNutrient system could

provide cost-effective nutrition while reducing waste. At least it beat Soviet cosmonaut Yuri Gagarin's 1961 toothpaste tubes of pureed beef and chocolate sauce.

Welcome to CUBES, based at the University of California, Berkeley. Under its voluble director, Adam Arkin, researchers sent the first BioNutrient packets to the International Space Station for the use of astronauts, who then froze the results for post-landing examination. Using a programmable 3-D-printer, the BioNutrient system ignites bioreactions to make food ingredients. The programmable device can receive new genetic instructions from Earth, forming recipes for yeast fermentation of new food flavors. This five-year experiment is allowing researchers to observe how much food engineered yeast can grow, as well as its quality of taste.

Crop engineering can also enable plants to grow without photosynthesis. For CUBES, Utah State researcher Bruce Bugbee engineered crops to convert carbon dioxide into food, without needing sunlight and while using less water than on Earth. Bugbee pointed out that on Earth, crops are the biggest wasters of our planet's freshwater, and crops that can be grown with little water and sunlight would benefit astronauts and Earthlings alike.

Imagine having a severe headache and reaching for lettuce as a cure. This was the project of another CUBES team member, University of California microbiologist Karen McDonald. Aiming to keep astronauts healthy without multiple pill bottles, she and her team hoped to produce space plants with medicinal benefits. Using CRISPR technology, McDonald and her team worked on development of lettuce possessing antibiotic or painkilling capabilities. Medicinal lettuce would not exclusively benefit space travelers. In an era of rising drug prices, it could provide medicine in neglected communities globally.

The biggest impediment for synthetic biology in space is the enormous price per pound of payload it costs to break Earth's gravity. Kate Adamala told me: "There's a steep learning curve when your things fly. I never thought too much about limits like power budget, weight, or atmospheric composition. I was going to use green fluorescent protein for our microbes and realized, 'Wait, we don't have enough oxygen for them to glow.'"

Mastery of that learning curve was a challenge occupying the unique, California-based NASA researcher Lynn Rothschild.

MAKE IT, DON'T TAKE IT

In the arid mountains around Mountain View, California, the rocky soil bakes in the summer heat. At the NASA Ames Research Center, Lynn Rothschild was talking with her team about biomining.[15] Astrobiologists have spent ten years developing matchbox-sized biomining reactors on Mars to use microbes' ability to extract minerals from rocks.[16] Eighteen of them went up to the International Space Station and performed admirably, forming microbial biofilms rich with minerals such as iron, calcium, and magnesium from rocks. But Rothschild was thinking bigger. Much as poets say that strict formulas make them more creative, Lynn Rothschild recommends a similar inspiration for space synthetic biology. She has directed ten Brown and Stanford University iGEM teams researching topics like biomining and fungi-made materials. When a 2018 student mentioned that fungi bind metals, something clicked. "I didn't even hear what he said after that," Rothschild remembered. "I know about fungi. We could attach a protein to the cell wall, I realized, and it would act as a water filter. It worked so well it was amazing."[17]

Long-duration missions require technologies for radiation protection, habitat repair, and various forms of life support. Rothschild sums up the challenges succinctly in her humorous, direct style. "You can't just say 'Oops, I forgot . . .'" The solution, according to her and like-minded scientists, is life. The use of life as a technology.

Synthetic biology is what may enable organisms to thrive in extreme environments to produce beneficial chemicals desired by astronauts. The potential for such genetic design was becoming more hopeful after synthetic biology successes on Earth. Scientific developments such as altering the genes of pests with CRISPR and building packaging, furniture, and housing with mycelia made some of the older plans for terraforming appear obsolete. Under Rothschild's enthusiastic direction, NASA was moving into the business of redesigning cells to be factories.

With her doctorate in cell and molecular biology from Brown and bachelors of science from Yale, the outgoing Rothschild is an avid researcher of life on other planets, a field known as astrobiology. She has studied extreme life in Kenya's Rift Valley, Bolivia's Andes, the Australian outback and in New Zealand's hot springs. She started with protozoa, one-celled organisms like amoebae, and then won a one-year contract at NASA to seek out life on other planets. It seemed like her dream position, even if only for what was originally supposed to be two years.

Late that year, she attended a thousand-person lecture by NASA director Dan Goldin at San Francisco's Moscone Center. As she waited for him to begin, she introduced herself to the program chief at NASA Ames Research Center, who stood in the VIP section, along with the agency's public affairs director. She talked to him about her work, and his ears perked up.

"Lynn," he asked. "Wait, do you mean you're a cell biologist?"

Well, yes, she said, that was her PhD thesis.

"Welcome," he said. "You're hired full-time!"

Rothschild went on to create the influential biannual Astro-biology Science Conference (AbSciCon) and to lead iGEM teams as synthetic biology became part of her search for alien life. Rothschild did so well she was named by biotech company Synthego as one of six critical women in synthetic biology in 2019.[18] Surrounded by excited undergraduates crowded into her lab every summer, she would get behind a microscope and lose herself in looking at her first sources of study, beautiful protozoa.

The biggest contribution synthetic biology can make to space travel is to solve the "upmass problem," Rothschild explains, echoing Kate Adamala, the enormous expense of lifting weight into space. Instead of hefting medicines and equipment at a price of $20,000 per kilogram, you could instead carry tiny, freeze-dried programmed microbes to make the chemicals and material you need. "What about making your clothes (or) recy-cling the clothes you're wearing?" Rothschild asks. "Making your own detergents? The materials of the spacecraft itself?"[19] Microbes are not only programmable, they are self-repairing and self-replicating. They could produce chemicals for products with high fidelity at incredibly small scales, all while surviving on minimal resources in the most hostile environments.

Those ideas were thrashed out among her regular lab members and in iGEM contributions from her student teams, several of which won NASA Institute for Advanced Innovative Concepts awards. Her team members called their extreme microbes "hell cells," putting them to work to make medicines and bricks for buildings. Rothschild declared to audiences: "Synthetic biology is the art of the possible. We must learn to make it, not take it."

The principle was to use self-sustaining organisms to recycle waste into materials, foods, medicines, and energy sources. You could transform coffee grounds into plastic plates, for example, and recycle waste into valuable products. NASA sponsored a Deep Space Food Challenge to engineer microbes to make ingredients for fresh foods. As new technologies on Earth progressed, updated DNA instructions could be sent from Earth to fermenters on Mars, using knowledge both new and old.

The Rothschild team program for the use of mycelia was to make habitats grow on other planets. Adding water to a lightweight, small, dormant fungus could gradually grow a domed structure. The top of the structure, placed outside with water, would become ice-covered. The ice could be slowly melted down, dripping onto an underlying layer of cyanobacteria, or pond slime, which could in turn photosynthesize faint sunlight to create oxygen for astronauts and food for more mycelia underneath.[20] Mycelia could also be used for biomining. In a 2019 *Nature* article, Rothschild and colleagues published a patented approach to using mycelia for water filtration.[21] With the patent in hand, her team turned to using microbes to extract minerals from and clean wastewater.

As for medicines, insulin would be easy to bioengineer from bacteria or yeast, as on Earth. Other treatments, like one to combat bone density loss due to low gravity, were also possible using fairly straightforward microbial manufacturing or, in the case of bone loss, planting calcium-building transgenic lettuce seeds.[22] Acetaminophen could be extracted from modified plants, much as aspirin was first derived from the bark of willows.

A colleague of Rothschild's, MIT astrobiologist Lisa Nip, went further. She explored the potential of synthetic biology to augment human survivability in deep space. "Synthetic biology will be a means for us to engineer not only our food . . . but also ourselves," she says in lectures. The resistance of the microbe *Deinococcus radiodurans* to high radiation is a model, she argues. Nip believed genetic modification could supplement our melanin's ability to protect us from radiation. The enthusiastic researcher relates such plans to the ways in which Tibetans evolved to survive on low oxygen and in which some Argentinians developed a tolerance for the arsenic-contaminated water of the Andes.[23]

Both Rothschild and Nip end their public talks by addressing fears associated with genetic editing. Rothschild challenges her audiences to consider that "No one in the room has a wolf cub at home, but you might have a puppy or a dog. . . . We've been genetically modifying organisms for tens of thousands of years and more," she reminds listeners.[24]

Back on Earth. NASA was testing biology to address another profound need of astronauts: their emotional well-being.

LIFE ON MARS

You are one of six people trapped in a metal box on the top of a volcano, on one of Earth's most frightening rock formations on the Big Island of Hawaii. You cannot leave for eight months. Although little in danger of the volcano's erupting, you will end up surviving one earthquake.

A technical biologist, a social media expert, an engineer, and an agronomist were among the group of twenty- and thirtysomethings with resumes ranging from Google to graduate school who were placed on top of a mountain in 2017 by NASA for its Hawaii Space Exploration Analog and Simulation (HISEAS) project. Their only contact with family, ground control, or anyone else "on Earth" was through e-mails featuring a forty-minute roundtrip time delay imposed by NASA. The team was living a simulation of life on Mars to learn what may keep a space crew happy and healthy during an extended mission.

The job of the participants was to turn themselves from strangers into a smoothly functioning team, all while tracking each other's food consumption, interpersonal dynamics, relaxation behaviors, moods, emotions, roles, and performance, along with other technical assignments.

This was the group that the goateed Floridian with a green thumb, Josh Ehrlich, joined from January until August 2017. His team was proving, among other tasks, that humans can engineer microbes to make proteins and chemicals they might need on long space flights. As mission specialist of biology, Ehrlich focused his efforts on studying plant growth, using a model called Veggie from an ongoing project on the International Space Station.[25]

Ehrlich was not only a green thumb but also an engineer. He had helped his single mom and grandmother plant their squash garden in Hollywood, South Florida, where he grew up. On the Hawaiian mountaintop, it was the unexpected findings that made HI-SEAS an ultimate life experience for Ehrlich. "I got really interesting results that came from the plant lighting, with the green LEDs, primarily," he informed me. "The neatest thing was understanding light variations that mimic those on Earth optimize the growth of a plant, also provide mood lighting for the crew!" Ehrlich was studying space engineering and, after HI-SEAS, went on to work for Lockheed, on swarms of miniature drones to find water on planets, and on various other robotic and human-rated spacecraft including Orion and Mars Base Camp, as well as spaceflight programs with an emphasis on Low Earth Orbit and the Moon.

His crewmates in Hawaii were a collection of dreamers. Laura Lark was a former Googler in charge of the NASA module's enhancement. Their talkative British science officer came from the European Astronaut Centre, and their captain,

James Bevington, was a Tennessee-raised astrobiologist working as a visiting researcher in Danielle Tullman-Ercek's Northwestern University lab.

Dropping everything to do this research for NASA was "extremely disruptive" to their lives, Bevington said. Most everyone "quit their job, or didn't know if they were going to have a job," he added. Yet the result was the most formative experience he would ever know. "Before I was very technically focused," Bevington recalled, "all about data, results, measurements, and statistics. I came to understand that people and teamwork are more important. If you don't have a functioning group, it doesn't matter how much knowledge you have."

Some of their research included synthetic biology. For astronauts assigned to Mars trips lasting years, needed medicines would pass their expiration dates, Bevington pointed out. "So you need a way to make them." With synthetic biology, you could program a plant to make aspirin, as other labs had proposed, and then eat the plant when your head hurts. It could not be any worse than college days of instant ramen.

Emotionally, life in space was not easy, however. The HI-SEAS program was judged successful but also dangerous, and when a subsequent mission had to be aborted because of one participant's illness, NASA suspended the program pending review. For his part, Josh Ehrlich moved on from Lockheed to Jeff Bezos' aerospace company Blue Origin, where as senior aerospace systems engineer, Ehrlich built a unique AI system for ground-to-space communication on long-duration missions, installed in the summer 2022 on NASA's Orion spacecraft, from there went to work on the Artemis mission.

Meanwhile, on our stricken planet, a much bigger and more dangerous crisis was continuing, despite the help of synthetic biology.

"BIOLOGY . . . EVOLVES TO SOLVE NEW PROBLEMS"

Ginkgo's cofounders rang the New York Stock Exchange's opening bell to inaugurate their company going public. "The future is to grow" proclaimed a company banner suspended over colorful sculptures of plastic flowers and DNA helices that hung from the Stock Exchange's façade. The launch raised some $1.6 billion in the biggest biotech IPO to date.[26] At a Times Square rally, Ginkgo distributed giveaways of the stock listing translated into a DNA sequence encased in orange plastic to evoke the amber of *Jurassic Park*. "Biology makes *stuff*, and it evolves to solve new problems," Ginkgo noted in its Wall Street Form S-4 announcing the public offering[27]

The launch occurred in a strong world economy. Labor was in short supply. Home prices were climbing almost beyond the means of many buyers. Things appeared to be looking up. But waves of new SARS-CoV-2 variants kept sweeping regions from the former East Germany in Europe to Texas in the United States, stoking fresh fears and renewing shutdowns. Inflation too was rising. Universities, government agencies, and major companies like Bayer and Merck were pursuing programs in fields that had raised billions of dollars. A €25 million ($28,201,000) Dutch collaboration of multiple universities was trying to build an artificial cell. Some researchers thought it might prove to be easier to make new life forms than to understand them. Historian Sophia Roosth commented that synthetic biologists tended to build novel life forms and then try to explain them.[28] The question seemed to change from *whether* synthetic biology would make an industrial revolution almost to *when*.

Synthetic Biology journal editor Jean Peccoud argued that the new science's economic contribution combined elements of manufacturing and computer programming. Abell Chair in Synthetic Biology and professor of biological engineering at Colorado State University, Peccoud noted that synthetic biology cell factories made products, but those cells featured the artificial intelligence techniques of the computer. Biology researchers no longer edited DNA themselves, but rather sent their desired sequence to companies like IDT or Twist to have the DNA made. Ginkgo offered cell development kits (CDKs) in imitation of programmers' game development kits (GDKs). "What the field needs," said Peccoud, "is a universal, standardized directory of interchangeable biological parts."[29]

On a crisp winter night in the Gore Mountains of Colorado, I stare at the sky, thinking about what Rothschild, Arkin, Peccoud, and Adamala had told me. Above me, Mars glowed red, seemingly close enough to touch. A few miles away, commercial and private jets at Eagle County Regional Airport are flown on fuel made in part from beef tallow that burns like the traditional fuel with which it is mixed. Here on Earth, the rapidly accelerating science of synthetic biology, combining metabolic engineering, standardized biological parts, gene editing, directed evolution, and semisynthetic organisms, seemed poised to go to the next level of producing ingredients for sustainable products. Scientists were no longer beholden to evolutionary constraints in engineering life. The time had come to judge what synthetic biology could do for all humans on Earth, not just the privileged like me.

15

FUTURAMA

Life is very short and what we have to do must be done in the now.

—Audre Lord

The crowd waited expectantly at the packed Oakland Marriott City Center. It was mid-April 2022, and the skies had cleared from a cold, windy rain in the California city. I walked through the gleaming exhibits of microbe-made sustainable skis, fungi-based building materials and pungent bacon, whole labs on a computer chip and green-glowing house plants, eager to hear about topics such as the "third agricultural revolution" and synthetic biology-derived medicines for mental illnesses. The founder of clearinghouse Synbiobeta, former astrobiologist John Cumbers, stands at the front of the stage in a black vest. Attending this conference were scientists such as Harvard's George Church, MyBacon's Eben Bayer, Twist Bioscience's Emily Leproust, Persephone's Stephanie Culler, and others I admired, like the iGEM competition administrative team and the local high school and college students supported by scholarships from Ginkgo Bioworks and the foundation created by former Google CEO Eric Schmidt.

Outside the hotel, the world remained in stress, with new SARS-CoV-2 variants requiring new versions of the successful vaccines. The stock and housing markets were gyrating as inflation, worker shortages, and fluctuating oil and gas prices pummeled economies. The flow of easy venture capital money of the previous years had slowed.

On the one hand, some synthetic biology companies were flying high: $17.6 billion had been invested in 2021, with much more predicted in 2022. Some forty-two publicly held companies were listed on the stock market. By fall 2022 the science would be the focus of a national plan that contained most every provision the field's leaders had wanted when they began sixteen years earlier.

The best way to judge synthetic biology is broadly, as that part of economy based on products, services and processes derived from engineering biological sources such as plants and microorganisms. By 2030, the global bioeconomy was projected to be worth between $4 trillion and $30 trillion.[1] COVID-19 vaccines showed how quickly biotechnology could produce world-saving medicines. Messenger RNA as a platform was being studied in labs all over the world. The fast-growing companies Twist and Ginkgo signed a four-year deal to provide engineered DNA to meet the needs of innovative start-ups. The White House wanted to invest in fermentation factories and markets for synthetic biology-made goods.[2] The potential applications of synthetic biology breakthroughs are wide and diverse.

On the other hand, many people could not afford products that remained scarce and high-priced. In South Africa, researchers sought to devise their own COVID-19 vaccine without paying Moderna and Pfizer pricing, seeking a decentralized mRNA platform to treat illnesses. Responding to pressure, Moderna and Pfizer were building manufacturing plants in Africa and in South Asia. Neither, however, was lowering its prices or sharing its formula. Some eighty patent fights raged around the world for control of vaccine development and distribution.³ The patent battles over CRISPR gene editing were continuing. Some of the outlandish synthetic biology forays seemed ill-considered, such as the idea of resurrecting the mammoth.

Globally, the technology was spreading. Indian Oil signed a deal with LanzaTech to set up the world's first refinery of waste-gas-to-ethanol production in Haryana, India. In the lake city of Kisumu in western Kenya, the Fifth Africa Biotechnology Conference concluded in November 2021. The month before, the Inaugural International Synthetic Biology Conference was held in Kampala, Uganda. iGEM had more than 350 teams from 40 countries compete at 2022's conference in Paris.⁴ Synbio-Africa in Entebbe, Uganda, featured an iGEM showcase.

In the United States, several institutes were trying to attract more diverse researchers into the field. Funded by the Department of Defense, the nonprofit educational organization called BioMade was sponsoring institutes on sixteen college campuses to draw in students. With support from the National Science Foundation, another nonprofit, called InnovATEBIO, was sponsoring high school, community college, and college programs at 134 campuses around the United States. In Cambridge, Massachusetts, the organization called BioBuilder brought together entrepreneurs, high school students, and researchers to provide bioeconomy student resources and training.

In cosmetics, clothing, and foods, products ranged from the high end to the middle of the road. The exclusive restaurant Silo London was incorporating biology-built materials in its décor. Its lightshades were made by Ecovative mushroom root technology. Its dinner plates were fashioned from recycled plastic bags and its tables from reconstituted food packaging. Stella McCartney's mycelium-made designs for Bolt Threads were among the designer's biggest sellers. Meat producers like Tyson and Cargill inaugurated plant-based divisions, and Cargill was building a sustainable manufacturing facility using Genomatica synthetic biology technology. Dairy producers Chobani and Dannon started milk-free divisions based on synthetic biology. The world's first butcher shop selling lab-made meat opened in Singapore. The Israeli company Believer Meats broke ground on the world's largest cultured meat factory in Wilson, North Carolina.[5] Not all those wonders seemed destined for mass consumption, however. Many lab-derived foods were less tasty than their natural counterparts. DNA-based computers had not materialized. Biomanufactured clothing served a small market.

Did the new science promise to become an industrial revolution? Was another "industrial revolution" even desirable? To answer such questions we start in the past, follow with current government policies, revisit a sustainable fuel manufacturer, meet a synthetic biologist and the themes of her science, and conclude by returning to where we began this book.

A SUSTAINABLE INFRASTRUCTURE

Three industrial revolutions created the world we live in.[6] The first, in the late 1700s, came from coal and the steam engine and led to the transformative ship-to-rail transportation connection, setting off the shift from agricultural to industrial economies. The second, in the 1870s, sprang from electricity and oil, producing the factory and automobile assembly line and the steel mechanization of the early 1900s. Industrial revolution number three, in the late 1960s and 1970s, encompassed silicon and nuclear power. To these many experts now add a fourth, our current revolution combining artificial intelligence and satellite communications, giving us the internet, social media, cell phones, and apps that make consumption so easy and so stressful. While economic historians may argue about the numbers and dates, each of these upheavals created new levels of wealth amidst transformed social structures. Industry took us off the farm and into the city.

The first three revolutions resulted from energy sources that inspired novel technologies. By contrast, the fourth revolution combines information-processing breakthroughs with satellite communications and miniaturization. The first three were directed outward, the current one is directed inward. If synthetic biology is to revolutionize industry, it will be an upheaval in biological production, dependent on more affordable prices and higher-quality products, uniting the lab techniques we have covered—metabolic engineering, standardized parts, gene editing, directed evolution, and semisynthetic organisms, with new ones from artificial intelligence and 3-D printing.

At its beginning, synthetic biology was a small community of researchers funded mostly by government grants. They applied the recombinant DNA triumphs of the 1970s and 1980s to engineer increasingly sophisticated gene circuits in cells. Some sought to transcend capitalism and erase traditional boundaries between professional and amateur and commerce and academics.

Government played a key role. When Jay Keasling's Berkeley achievement in 2006 made a malaria drug available to people in need, it and many subsequent efforts were funded by federal programs sponsored by DARPA, the National Science Foundation, and the Departments of Energy and Defense. Companies as different as Moderna and Ginkgo won their seed money from such programs, in which a few government officers foresaw the threats of biosecurity, climate change, and new viruses. They pushed academics to devise solutions. The National Science

Foundation funded the biannual gatherings of the Synthetic Biology Engineering Research Consortium (SynBERC) and, when its grant ended, underwrote the Engineering Biology Research Consortium (EBRC) to drive the field forward. The publicly funded Joint BioEnergy Institute (JBEI) in Emeryville, California, was where many companies got their first access to lab equipment.

Investment of Silicon Valley money followed and, with that, the push for profit. The Gates Foundation funded the malaria medicine success. Most every company's initial public offering was fueled by venture capitalists seeking a windfall. The venture capital firm DCVC was a big investor in Ginkgo.[7] Other investors included London-based Atomico, where partner Siraj Khaliq gushed about the promise of synthetic biology, saying, "This is the most important technological revolution of our time." Cathie Wood, venture investor in California who made a fortune on her early Tesla bet, told audiences, "The future of investing is investing in the future." Such pronouncements glossed over company failures.

Government grants plus venture capital sustained the field from 2000 to the present. Virtually no profit yet came from consumer or wholesale sales. The small successes were mainly in consumer products such as ingredients for flavors and fragrances, as well as fashion and packing materials, jet fuels and medicines, and one world-saving vaccine. In foods, the plant-based meat platform was expanding from the luxury market to fast foods and cheaper restaurants. The brand Peet's Coffee, for instance, produced plant-based breakfast sandwiches made by the company EatJust. In medicine, messenger RNA promised to become a platform of new therapeutics and vaccines. In agriculture, Pivot Bio's fertilizer replacement was on pace to cover 5 million acres' worth of land. Others engineered microbes to free farmers from the use of pesticides. In medicine, potentially therapeutic DNA was being 3D-printed.[8]

Around the world, several government programs tried to jumpstart the industry with policy initiatives. Here are some that worked.

SUCCESSFUL GOVERNMENT POLICIES

Synthetic biology was embraced more readily, and sooner, by governments other than the United States. Some countries needed to maximize agricultural output. Singapore, Israel, and the Netherlands became leaders in making cell-free food ingredients—proteins created in the lab on natural scaffolds—followed closely by the UK, Spain, and Germany. Singapore invested $20 million in the Synthetic Biology for Clinical and Technological Innovation (SynCTI) program at its National University and cofounded the Global Biofoundry Alliance. The Israeli government funded cell-free meat manufacturing by companies such as Rehovot-based Aleph. Australia's national science board became a leader in sponsoring biomining. In the United Kingdom, the leading institutes include Oxford and Cambridge Universities, the University of Edinburgh, and the Warwick Synthetic Biology Research Center.[9] Imperial College London's Synthetic Biology Hub features three branches—the Centre for Synthetic Biology, SynbiCITE, and the London DNA Foundry. In Germany, experts drafted recommendations to support research led by the prestigious Max Planck Society. The authors of a 2022 Schmidt Futures Report criticized the fact that many U.S. synthetic biology companies turned to "manufacturers in Belgium, Canada, China, Germany, India, Mexico, the Netherlands, Slovakia, Slovenia, and elsewhere" for bioproduction, calling for a policy to boost U.S. fermentation infrastructure.[10] The same call for more such facilities was echoed in Europe.

The Chinese government was a world-leading supporter of the science, sponsoring some nine synthetic biology projects, including state-controlled fermenters. In Hong Kong, the plastics and real estate magnate Li Ka Shang made the biggest single private investment, contributing $63 million to a synthetic biology research center.[11] The tech hubs of Shenzhen and Beijing housed two institutes with ties to well-known researchers, Berkeley's Jay Keasling at the Shenzhen Institutes of Advanced Technologies (SIAS) and Harvard's George Church at the Beijing Genomics Institute (BGI). At the end of 2020, China announced the creation of a joint government research center with Belgium.

In the United States, the EBRC had since 2007 brought together academics and entrepreneurs to shape policies calling for federal investment in bioproduction. It proposed a series of regional fermentation centers, as had been done with wartime penicillin production in the twentieth century. Experts called for an innovative approach to regulatory policies, some of which were contradictory. Researchers called for money to train a diverse workforce in the new jobs.

That policy effort has succeeded. The late 2022 White House Executive Order on Advancing Biotechnology and Biomanufacturing Innovation made synthetic biology the focus of a sweeping national industrial program. "The COVID pandemic has demonstrated" the report began, "the vital role of biotechnology and biomanufacturing in developing lifesaving diagnostics, therapeutics, and vaccines." The order called for $2 billion to expand the infrastructure of biomanufacturing—meaning fermenters on the scale of those used to make the COVID vaccines–and to coordinate government policies amongst different divisions including Health and Human Services, National Security, and the Office of Science and Technology.

The plan envisions better coordination of policies regulating human-made biological entities. In the United States, the oversight of synthetic biology products is divided among three different federal agencies, the Food and Drug Administration (FDA) for food and drugs, the United States Department of Agriculture (USDA) for remade food crops and livestock, and the Environmental Protection Agency (EPA) for organisms that might assist in industrial remediation. Such regulation must be streamlined for this science to make a thriving industry.

Next in the plan is to train a bigger workforce, not only in the biotechnology centers of Boston, San Francisco, New York, Illinois, Iowa and Texas, but in other states as well. Additionally, some of the same agricultural states could be tapped as providers of biomass straight from the farm, slaughterhouse and ranch.

Finally, the plan aims to support markets through subsidies to trucking companies and auto and boat makers to adopt the products of biotechnology. Mercedes, for instance, featured Mylo mushroom-made leather in some cars, as we have seen. Noting the need for the United States to remain a global leader of innovation, the plan envisions a nation powered by biofuels, treated by biomedicines, clothed and housed in sustainable bio-materials, readily available at affordable prices.[12]

Still unclear is whether the various programs will be implemented and, if so, how well whether they will work in the marketplace.

One model of how synthetic biology could sell products remained that of the California company Amyris, originally funded by the Department of Energy and DARPA. Amyris had its own factory and introduced cosmetic ingredients such as squalane, a synthetic yeast-made version of shark oil squalene. Its

Biossance cosmetics and Purecane sugar substitute brands were expected to top $300 million in sales in 2022. This is a traditional way to make money as a manufacturer. "When we sell a kilo of squalane directly to the consumer," John Melo told *Barron's*, "we get $2,500 per kilo. When I sell it to another beauty company, I get $30. Do the math."[13] Analysts projected the company could be profitable by 2024, a first in the industry.[14]

Another model was that of Massachusetts-based Ginkgo Bioworks, a platform company in something like the way Apple provided a platform for other app makers. Ginkgo reported $478 million in revenue in 2022, representing an increase of 52% over 2021. It was the world's largest business-to-business synthetic biology company, with a seventh Bioworks completed in Emeryville, California, and expanded partnerships with companies like Lygos, Synlogic and Bolt Threads, to optimize their biomanufacturing. The organism was their product, as the founders envisioned. Ginkgo was also expanding its partnership with the German corporate giant Bayer to apply synthetic biology to agricultural products. It developed biosecurity technology to detect if a pathogen is a product of human engineering of interest to law enforcement.

Thus far, synthetic biology businesses have offered a series of modest successes. "Amyris is a solid base hit," Bolt Threads' Dan Widmaier confided."[15] Many of the popular synthetic biology products remained expensive and limited in availability. The North Face ski jacket made by the Japanese company Spiber costs more than jackets by its natural-fabric competitors. In food production, the capacity of fermentation tanks remains far below what would be necessary to replace meat.[16] MyBacon's Eben Bayer argued for new biological factories, such as engineered moss, to make ingredients.

Other backers looked to industrial goods. DuPont was finishing a state-of-the-art Industrial Biosciences complex in the

Netherlands. Cities in Asia recycled carbon dioxide and methane into fuels. Some of the goals from the EBRC had included enabling nature to increase carbon uptake, engineering crops to more efficiently assimilate nitrogen, making biofuels more efficient, manufacturing better bioplastics, and reducing the footprint of materials and industrial and chemical manufacturing processes, but none of these was happening on a commercial scale.

Having spent so much time with academic scientists, I realized if I were to answer whether synthetic biology would ever constitute an industrial revolution, I had to return to a commercial setting. I scheduled a second visit to jet fuel and chemicals manufacturer LanzaTech.

POLLUTION TO PRODUCTS

On a last warm morning of November 2021, I returned to LanzaTech in Skokie, Illinois.[17] The parking structure was much fuller than it had been two years earlier. "We've doubled in size," Michael Köpke, vice president of synthetic biology, told me. LanzaTech was making sustainable aviation fuels (SAFs) for several airlines and many new consumer products ranging from cleaners to clothing in partnerships around the world. It employed more than 200 people working in facilities in China, the United Kingdom, India, and the United States. Its Skokie labs featured divisions for proteomics (the study of proteins), metabolomics (the study of metabolism), and artificial intelligence processing. The fermenting room tested the company's designed organisms in thirty-meter-tall steel tanks.

Köpke guided me to put on my carbon monoxide detector and goggles. Dressed casually in jeans and a gray hoodie, he remained an aficionado of Chicago's reggae clubs. As we discussed the company's special project being pursued in association with Northwestern University's Danielle Tullman-Ercek to repurpose bacterial microfactories to produce synthetic rubber, funded by the Department of Energy, I recalled my first visit to Tullman-Ercek and her work on promising drug delivery vehicles in 2019.

Our LanzaTech tour began with its consumer product exhibits: Swiss "Potz" cleaning supplies where both ingredients and packaging are made from LanzaTech microbe-made alcohols, waste carpet that LanzaTech can convert to new products, and the biology-made material of Lululemon sports apparel. Manufacturer of one hundred new industrial molecules, including different kinds of ethanol, isopropyl alcohol, acetone, and others, the company displays a stock market–type LED ticker on one wall showing how many thousands of tons of carbon dioxide its factories have removed from the environment. The reading was 150,000.

LanzaTech now had two large factories in China (the original near Beijing and a new one near the Mongolian border) in operation and others under construction in India, the United Kingdom, Canada, and at Belgium's Arcelor Mittal steel factory in Ghent. Other facilities include a Japanese landfill site to pilot its ethanol-producing recycling process from waste. In Japan, the company was creating bioplastics from gasified trash. A larger factory is currently under construction in Soperton, Georgia, again targeting ethanol for jet fuel but with plans to start production of acetone and other chemical ingredients. The Soperton factory featured a test site of new microbial strains shipped to other production facilities.

Recent activities included an arrangement with SAS Airways, Shell, and the Swedish power company Vattenfall to build a sustainable jet fuel factory in Europe and produce some 50,000 tons of fuel a year from waste gases. The Swiss sports clothing giant On signed a deal with LanzaTech as did renewable plastics leader Borealis, the latter to make the soles and sides of On's popular sneakers.

After a tour of the mixers, shakers, reactors, and anaerobic solid glass compartments of the reactors, Köpke introduced me to chief sustainability officer Freya Burton. "Brands come to us," she said, referring to the company's technology that can convert industrial waste gas by fermentation into products like fibers for clothing material. "We are creating industrial symbioses. Who would have thought that our steel mill partners would end up working, through us, with Lululemon or L'Oréal?"

LanzaTech had also signed an agreement with India's government to convert agricultural waste into ethanol for transport use. Expecting to become the world's third largest aviation fuel maker by 2024, India is a key market for LanzaTech's alcohol-to-jet technology. Much as with steel waste gases, the technology converts the discarded stalks, leaves, and branches of farm waste into fuels, textiles, detergents, soaps, and packaging materials. To promote such collaborations, U.S. secretary of energy Jennifer Granholm and Georgia senator Jon Ossoff visited LanzaTech's Freedom Pines manufacturing plant in Soperton, Georgia, in October 2021. Utilizing waste carbon oxides from a variety of feedstocks, that facility when completed is expected to churn out 10 million gallons of sustainable aviation fuel within five years. British Airways, Alaska Airlines, and All Nippon Airways have committed to buying the jet fuel from the Georgia plant. Other initiatives included a European Union grant for its carbon capture technology and expansion of its clothing, fragrances, and packaging partnerships. In February 2023, LanzaTech went public.

As we walked through the white-walled labs in Skokie, Köpke talked about why he came to the firm from his native Germany. "A lot of these techniques we had to develop ourselves," he explained. "I believed in what we were doing." LanzaTech had tripled sales in one year. It had seven plants in development, with a third Chinese plant under construction.[18] As we met up with company founder Sean Simpson for lunch at a Mexican restaurant, Simpson answered the question then on my mind: "Definitely this will be an industrial revolution. We have to succeed. We have to see Ginkgo succeed, and others like us. The world needs us," Simpson said, as we munched on our enchiladas and taco salads.

Outside the windows, a weak sun shone over the quiet street. The new science, I thought, offered to change our relation to ourselves, each other, and the world. Driving back from Skokie, watching families out walking and bike riding, I reflected on how the new field was also changing what a scientist is.

WHAT A SCIENTIST IS

From its beginnings in the early 2000s at MIT, synthetic biology has fashioned itself as an inclusive science featuring community labs, DIY experimenters, and a commitment to understanding the ethics of changing life. Over the years, the science had opened its doors to different types of researchers. Many of the leaders of synthetic biology are women—Danielle Tullman-Ercek, Emily Leproust, Aoife Brennan, Christina Smolke, Kristala Prather, Tara Deans, LanzaTech's CEO Jennifer Holmgren, Steffanie Strathdee, not to mention Nobel laureates Frances Arnold, Jennifer Doudna, and Emmanuelle Charpentier. Others came from the DIY community. Emory

University's Karmella Haynes argued that synthetic biologists did not necessarily need PhDs. Molecular biologists such as Yale's Natalie Kofler became involved in educating young people about synthetic biology ethics by maintaining the website www .editingnature.org and co-teaching a gene-drive ethics course through the Harvard Citizen Science Initiative.

We have met the University of Utah's Tara Deans, associate professor in biomedical engineering and inventor. Whenever someone told Deans something could not be done, she was motivated to do it. With a mother who was a diplomat, the blue-eyed Deans grew up all over the world, including Germany where she witnessed the fall of the Berlin Wall. For her doctorate, she won a position in the Boston University lab of Jim Collins, where she was building a switch to turn off mammalian genes. When she asked Collins to try a new tack, coupling her repressor proteins with RNA interference to turn off the genes, he gave her two months. Her seminal experiment on changing mammalian gene expression appeared in *Cell* in 2007.

Deans was in an unusual situation in graduate school because she was the first female PhD student to have a baby. Accused by some senior faculty members of not taking her scientific commitment seriously, she gained the support of Collins and others to write the department leave policy for pregnant students. When she became pregnant a second time, however, she felt overwhelmed. Collins reassured her. "This is great news," he said. "How can I help?" He also insisted they patent her genetic switch published in *Cell*.

"I am a scientist," Deans recalled telling him. "I don't profit from my work."[19]

"No, you need to capitalize on this," Collins insisted. "This is your way of protecting your intellectual property." With that, she became an inventor. Some fifteen years later, when the rights to her genetic switch were bought, Collins's advice paid off with a welcome check for the associate professor raising a family on two academics' salaries.

The science opened possibilities for creative research, and young people, and conventional scientists to be celebrated in popular and high culture. Oxford's vaccine maker Sarah Gilbert had a Barbie doll based on her distinctive red hair and was named Dame Commander of the British Empire. BioNTech's Katalin Karikó was a role model to her former students and named one of four *Time* Magazine Heroes of the Year in 2021. The iGEM student competition was such an engine for commercial success that NASA and other organizations were copying its format. Companies like Ginkgo, Opentrons, Eligo Bioscience, and BluePha in China originated with iGEM. In its Lunar Gateway Project, "I've never experienced anything like this before," an award-winning Marburg, Germany, iGEM team member tweeted of her synthetic biology competition experience.

Community researchers played a big role as well, like those in Brooklyn's Genspace lab. The DIY community of researchers contributed to the new science, and the diversity of that community is still growing. A beguiling story remained that of Opentrons, whose cheap robots helped New York City respond to the pandemic and were lifesavers in financially strapped labs and hospitals in Europe and Africa. Many other synthetic biologists led businesses—Kristy Hawkins, Stephanie Culler, Aoife Brennan, and MIT professor Kristala Prather, who sat on the board of Inscripta, the digital genome engineering company, and ran the business Kalion with her husband, Darcy.

In a final group were the artists and designers, from Stella McCartney to the artist Alexandra Daisy Ginsberg, lead author of the book *Synthetic Aesthetics: Investigating Synthetic Biology's Designs on Nature*. Ginkgo published the company magazine *Grow* as part of its effort to help prepare people to think deeply about the changes to come, as an echo of the magazine *Think* that IBM published in the 1930s to explain to people what a computer was. *Grow*'s fall 2021 issue titled "Equity" included an essay by Sophia Roosth on a moment when the Black Panthers aligned with doctors to promote public health in Black communities.[20] "I don't think we've fully thought about how it will change our lives," BioDesign founder Daniel Grushkin told me. I followed its breaking news in such Twitter handles as Gen News, I Cloned DNA, DIY Bio, Girls Who Code, and IFL Science (I F—ing Love Science).

"Synthetic biology is not just changing *E. coli*," Christina Agapakis told me. "It's changing people. We need to do the social work with students and teachers to make road maps for shaping politics. We need a parallel track in education about the ethics of this science."[21]

Part of that effort is to consider the principles of altering life and profiting from it. DARPA's Safe Genes program required researchers, for instance, to include a meeting with an ethicist. In May 2021, the EBRC published a document called "Guiding Ethical Principles in Synthetic Biology Research" that declared six rules—to create products that benefit people, to consider potential harm, to incorporate equity and justice, to openly distribute research results, to protect the rights of individuals, and to foster communication among stakeholders and the public.[22] With those principles in mind, the question became, How might this new technology change our place in nature? To answer that, we must look at its conceptual breakthroughs.

FIVE THEMES OF A NEW SCIENCE

This science depended on five conceptual breakthroughs. First, life is technology, and the principles of engineering can be applied to biological entities. Researchers can swap out cell controls, and the engineered organisms can produce ingredients for products. Second, the tensions between private and public, government, university, and business are becoming obsolete. You need all of them to innovate. Third, a combination of pure research and applied manufacturing is the path forward. Fourth, the future is interdisciplinary. Physicists, chemists, biologists, engineers, and computer scientists will all contribute and everyone will need to have access to its breakthroughs.

On the question of access, a series of initiatives sought to make vaccines available from regional sources of manufacturing less expensively than those of producers based in the United States and Germany. The World Health Organization banded together with South African universities and pharmaceutical companies in some fifteen countries to create an mRNA vaccine manufacturing center called Afrigen Biologics based in Cape Town, using nonproprietary technology.[23]

A fifth quality of synthetic biology manufacturing is that it can be local, on a small scale, facilitating access to energy, sustainable products, and medicines by more people. Places like Kaffee Bueno in Copenhagen were taking waste and making it into products, and Xilinat in Mexico City and Agricycle in Milwaukee were working to enable people to use synthetic biology for profitable waste recycling. Toward that end, the field's leaders "have to figure out how to design its infrastructure, governance, and policy components," said Michael Jewett, professor of chemical and biological engineering at Northwestern University. "We need educational programs and a global strategy to enable people to flourish, to create local bio manufacturing of medicines and foods."[24] When Amyris struggled with its giant Brazil refinery, it created a new one with five smaller bioreactors and much more efficient and reliable production.

The cell phone provides a model of how synthetic biology may create a revolution, said Jewett. The cell phone changed people's access to information processing. "What I hold in my hand is available across the planet. Cell phones changed the ways distributed computing can occur. In a similar way, synthetic biology has the potential to promote distributed biology for local solutions to global problems," Jewett said.

The most important challenge—local or global—would be to sustain Earth. The techniques covered in this book—metabolic engineering, standardized parts, gene editing, directed evolution, and new forms of genetic material—and their applications in medicine, the environment, clothing, food, housing, biofuels, defense, remediation, and biomining—promise a better future. First on the list is medicine, where synthetic biology may help confront bacterial resistance, and engineered microbes could help protect the body as sentinels and delivery vehicles, not to mention expanding their platform as mRNA and DNA vaccines. Next is energy, where engineered microbes turn carbon waste gases into aviation fuels or manufacture chemicals more sustainably than from current fossil fuel sources. Then there is remediation. Biomining and biorefining using synthesized microbes could reduce emissions and waste. Bioengineered microbes can draw down carbon from the atmosphere.

Synthetic biology offers the potential to be something different from the first three industrial revolutions. Those revolutions, coal, petroleum, electricity and nuclear, gave us enormous material gains but also the environmental crises we face today. Perhaps a new industrial revolution is *not* what we need.

The pandemic showed that if several billion dollars are spent, mass production breakthroughs in fermenting by designed bacteria were indeed quite possible in a short time. Such technical expertise can be extended. As of today, synthetic biology has created a raft of companies, numbering around 600, and the pace of innovation is expanding rapidly. The products include medicines like engineered CAR-T and stem cells to protect our health. The cancer-fighting CAR-T cell companies include Pfizer, Kite, and Allogene. For stem cell therapies, there are Vertex Pharmaceuticals in Boston and LocateBio in London, England. In cosmetics, there are Amyris, Genomatica, and Arcaea. For sustainable food and clothing, there are Huue, Bolt Threads, Spiber, Impossible Burger, Motif Foods, and Aleph in Israel. In remediation, the company Allonia engineered microbes to recycle municipal and agricultural waste. You can send a gene sequence to a genome company like Inscripta, and they will send you back new *E. coli* or yeast that you created.

This is not yet an industrial revolution, but it could be. Engineered biology will be part of a matrix of processes to create a more sustainable world. It offers a possibility of realigning the human relationship with Earth. Call it a production platform and a portal to a new society that manufactured in partnership with nature, not in opposition to or domination of it. The creation of new life forms could enhance the world for the improvement of human beings. With that thought, I decided to return to where I began.

REAL FARM

Outside the Oakland hotel atrium, I stare at a single red gladiolus, glowing with a color brighter than any human-made dye. Nothing man-made can compare to it. At the three-day-long Oakland conference, business owners have explained how they would use cell components the way electrical engineers used resistors, capacitors, and switches to implement a controlled output—except they would program cells using DNA, RNA, and proteins. Conference sessions have covered new ways of making drugs, cancer treatments, fertilizers, fuels and food, cosmetics and clothing. A human kidney grown in a genetically engineered pig kept a young person alive. In the area of biosecurity, synthetic biology is taking a lead in new virus prevention and detection. Some of the most widely used school COVID-19 tests are the products of synthetic biology.

Leaves were sprouting on the trees on Oakland's Broadway, with its marijuana dispensaries, police administration building, and homeless camp under the freeway. I stepped out from the show's high-glitz promotions, passing the murals honoring great Black artists—Jimi Hendrix, Tarika Lewis, and Ishmael Reed, to name a few who lived or performed in the city—and took an Uber to nearby Emeryville and the JBEI lab of Jay Keasling. There in the plant room with *Arabidopsis* and other green crops flourishing under blue lights, I had begun my journey four years earlier.

Keasling's father was retiring from farming after sixty years, he told me, and Keasling was going to give the commencement address at his alma mater, the University of Nebraska in Lincoln. I waited for a while before he raced into his office and motioned me to a chair. He unwrapped a turkey sandwich and told me about his new company, Zero Acre Farms, to produce sustainable cooking oils to replace the ecologically disastrous palm oil. The new product came out in July 2022.

He was moving as fast as ever in his field of metabolic engineering. Downing his sandwich ahead of a Zoom meeting with federal officials, he told me the world was at an inflection point, and synthetic biology was one of the best solutions to address the problems of climate and energy. He also had another partnership with GlaxoSmithKline to have yeast make a molecule for a health-care product. "We'll probably be able to talk about it more in a year," he said.

But what the lab was spending most of its time on was the engineering of large enzymes called polyketide synthases, from microbes, to make ingredients for use in sustainable plastics. But that was not all. The lab was partnering with a small Copenhagen lab to engineer yeast to produce a molecule for an anticancer drug. "It's probably the longest metabolic pathway that's ever been taken. It's from a roadside plant in California, and put into yeast," he told me.

From there, he shared his vision for the future. "I'd like us to be able to design biology on a computer and build it with robotics. I'd like to see a lot more companies that are cash-positive. We just talked about the ones that I'm founder of, but then there's a whole slew of companies that have come out of the lab." He went on to name yeastless beer-maker Berkeley Yeast, CBD maker Demetrix, Amyris, "and there's another company just a few blocks over here called Ansa. The two graduate students, while they were in my lab, developed an enzyme that would synthesize DNA without a template. And so it's an alternative to solid-phase DNA synthesis, which is the basis for all of the reagents we use now in the lab. To have that change is pretty amazing."

The clock was ticking. He had to get on the Zoom call to discuss biosecurity. As I walked out onto the sunlit industrial side street, I stopped at my favorite Mexican restaurant to jot down my notes. I recalled watching an earlier Zoom conference featuring Keasling. When he had explained in a conference session that he had grown up on a farm in Harvard, Nebraska, a young researcher in the audience had asked him: "What? A real farm?"

I thought of the synthetic biology horizon of new products: virus-sensing fabrics, cell-free systems to make antibodies and meat and fish products, and engineered microbes to make sustainable clothing, fragrances, painkillers, household items, and agricultural and industrial goods. I felt a mix of excitement and urgency in the idea to use biology as a technology. The farmer is an optimist, native American Will Rogers once said, "or he wouldn't be a farmer."

"Yes, a real farm," Keasling said.[25]

ACKNOWLEDGMENTS

I am thankful for the expert assistance and patience of many scientists who took time away from their work to share their ideas with me. The following are among those who stand out as helpful beyond the call of duty: Jay Keasling, Joshua Leonard, Danielle Tullman-Ercek, Julius Lucks, Michael Elowitz, Jim Collins, Tara Deans, Kristala Prather, Graham Hatfull, Lynn Rothschild, Mo Khalil, Leonardo Morsut, Kate Adamala, Wendell Lim, Michael Jewett, Chris Voigt, Adam Arkin, Tim Gardner, Zachary Serber, Aoife Brennan, Stephanie Culler, Emily Hartman, India Hook-Bernard, Christina Agapakis, and Steffanie Strathdee. Many researchers and communicators responded to my multiple queries over the years, including Pamela Silver, Rob Carlson, Steven Benner, Jonathan Glass, Ron Weiss, Reshma Shetty, Lynn Rothschild, Mamadou Coulibaly, Paul Turner, George Church, James Diggans, Jacqueline Fidanza, Elizabeth Vitalis, Daniel Grushkin, Meredith Fensom, John Cumbers, Alekos Simoni, and others.

Several researchers read and commented on my passages about their work. I am grateful to them for their time and extra effort. Commenters on portions of the manuscript included Emily Hartman, Jim Collins, Mo Khalil, Stephanie Culler,

Danielle Tullman-Ercek, Michael Köpke, Kate Adamala, Michael Elowitz, Graham Hatfull, Morgane Danielou, Joshua Ehrlich, Aoife Brennan, Chris Voigt, and Jay Keasling, and more.

Among friends and colleagues who read all or portions of this book as a manuscript, I acknowledge the excellent and generous intelligence of Bernie Selling, Michael Ladisch, and Howard Levi. Others who read parts of the manuscript include Laura Merwin and my DePaul University colleague and friend Stan Cohn. I acknowledge the thoughtful work of four anonymous peer reviewers who responded to the call of Columbia University Press. This time-consuming sharing of professional expertise is vital to writers of popular science like me.

In writing the book, I worked with several excellent DePaul University undergraduate and graduate students and alumni as research and writing assistants. Among these were Caitlin Howland, Rachel Latsos, and editors Miranda Lukatch and Mia Goulart. They helped conduct background research, wrote summaries, checked citations, and commented on passages. As a professional proofreader and fact checker, Ashley Belanger was an expert contributor. Another student assistant who contributed her research skills was Helena Petrouleas.

DePaul University provided direct support in four ways. First was a paid sabbatical to begin writing this book. Second was a critical Humanities Center Fellowship, which gave me more time to research and present portions of the manuscript to a group of supportive readers. Third was a Summer Research Grant to continue writing and editing, and fourth was a Competitive Research Grant to help pay for fact checking and indexing. In addition, the university's Office of Sponsored Programs provided excellent proofreading by the talented and enthusiastic Linda Levendusky.

At Columbia University Press, editor Miranda Martin was an expert supporter and guide. Brian C. Smith provided critical assistance with production and marketing.

Over the years, I often turned to the expert reader and sounding board, my wife Maja.

To these people, I am grateful. They allowed me to see this work from a different perspective as they provided their expertise. They contributed in a spirit of generosity and helped shape the best of these pages. As for any lapses, they are purely my own.

TIMELINE

10,000 BCE	Fermentation of wine, beer, cheese; domestication of dogs, cattle, horses
(Approx.)	Maize, potatoes, and other agricultural crops harvested
1818 CE	Mary Shelley publishes *Frankenstein*
1915	Chaim Weizmann's *Clostridium* bacteria produce a chemical used to make smokeless gunpowder
1970s	Improvements in polymerase chain reaction (PCR), DNA sequencing, and genetic engineering
1978	Genentech makes human insulin in bacteria and yeast
2000	Genetic circuit in *E. coli*, designed at Boston University and Princeton
2001	Frances Arnold, Caltech, mutates enzymes
2003	Tom Knight and others, MIT, develop the Bio-Bricks repository DARPA eighteen-month study of synthetic biology involving seven workshops
2004	Synthetic Biology (SynBio) 1.0 conference held at MIT
	First iGEM student competition held at MIT

2005	LanzaTech Biofuels founded in New Zealand
2006	Berkeley's Jay Keasling makes malaria drug artemisinin from bacteria
	Synthetic Biology Engineering Research Consortium (SynBERC) founded SynBio 2.0 conference held in Berkeley, California
2007	Joint BioEnergy Institute (JBEI) founded
	SynBio 3.0 conference held in Zurich, Switzerland
2008	SynBio 4.0 conference held in Hong Kong
	BioNTech founded in Mainz, Germany
2009	Bolt Threads founded in Emeryville, California
	Ginkgo Bioworks founded in Boston, Massachusetts
2010	Synthetic Genomics (J. Craig Venter Institute) creates Synthia 1.0, a minimal synthetic cell with about 900 genes
	ModeRNA founded in Cambridge, Massachusetts
2012	CRISPR gene editing developed by Berkeley's Jennifer Doudna and Max Planck Society's Emmanuelle Charpentier
2013	Twist Bioscience founded in South San Francisco
2014	Impossible Foods' Impossible Burger improves product and production
2015	Antheia, Cronos, and others make nonaddictive synthetic painkillers
2016	Bio-clothing and biomaterials increase in production
	Synthetic Genomics (J. Craig Venter Institute) creates Synthia 3.0, a more minimal synthetic bacterium of 473 genes
2017	Meeting at Berkeley on the ethics of gene editing

2019   Eight-base DNA, *hachimoji*, developed by
FfAME's Steven Benner
Ginkgo Bioworks resurrects scent of extinct Maui
mountain flower Microsoft, Twist Bioscience, and
DNA Bioeconomy use DNA as a computer
Twin babies subjected to embryonic CRISPR gene
editing by He Jiankui are born

2020   SARS-CoV-2 genome published on social media
CRISPR-based gene therapy heals sickle cell
patient Victoria Gray
Several synthetic biology COVID-19 vaccines show
promise Moderna and Pfizer mRNA COVID-19
vaccine approval and global distribution

2021   Ginkgo Bioworks, after going public, is valued at
$15 billion

2022   LanzaTech makes multiple national airline deals to
supply synthetic aviation fuels
Mycelium or mushroom-based clothing, packing
materials, and foods increase in production Biotech
estimated at 5 percent of U.S. GDP White House
Executive Order on Advancing Biomanufacturing
Innovation for a Sustainable, Safe and Secure
American Bioeconomy

GLOSSARY

ANTIBODY: A large, Y-shaped protein used by the immune system to identify and neutralize pathogenic bacteria and viruses.

BIOBASED MATERIAL: Material derived from or relying on biomass (in whole or in part).

BIOFUEL: Fuel that is derived from biomass.

BIOMATERIAL: Any biological substance that has been engineered to interact with biological systems for nonbiological use.

BIOMINING: The process of using microorganisms to extract economically valuable materials from rock ores, mining waste, or other solid materials (e.g., electronic waste).

BIOSENSOR: A device that uses in whole or in part living organism(s) or biological molecules to detect the presence of chemicals.

CARBON FIXATION: The process by which biological organisms convert inorganic carbon into organic compounds.

CAR-T CELL: Semisynthetic or modified immune cells to detect and attack cancers.

CELL-FREE SYSTEM: A synthetic biological system that activates biological reactions without the environment of a living cell. A cell-free system is an engineering biology tool for more controlled study of cellular reactions; simplified production of desired chemicals, biomolecules, or materials; or production in extreme or non-natural environments or with non-natural precursors or components. Cell-free expression is used for making proteins outside of living cells.

CHASSIS: An organism that serves as a foundation to physically house genetic components and supports them by providing the resources to function, such as transcription and translation machinery.

DIRECTED EVOLUTION: A process of speeding up and guiding mutation in the lab to explore new design space in organism engineering beyond natural selection.

DNA (*DEOXYRIBONUCLEIC ACID*): The double-stranded molecule that carries genetic information.

ESCHERICHIA COLI (E. COLI): A diverse group of bacteria found in the environment, foods, and the intestines of people and animals.

GENE EDITING: Changing DNA by changing the composition and sequence of nucleotides in a cell (aka *genetic engineering*).

GENE EXPRESSION: The process in which a gene's information is put to use in making a protein.

FERMENTATION: A process in which yeast break down sugar and make carbon dioxide and alcohol.

METABOLIC ENGINEERING: The practice of optimizing genetic and regulatory processes in cells to increase the production of certain substances.

PHOTOSYNTHETIC CAPACITY: A measure of the maximum rate at which leaves are able to fix carbon during photosynthesis.

POLYMERASE CHAIN REACTION (PCR): A method widely used to rapidly make billions of copies of a DNA sample, amplifying it for study in detail.

POLYPEPTIDE: A linear organic polymer consisting of a large number of amino-acid residues bonded together in a chain, forming part of (or the whole of) a protein molecule.

REPLICATION: The process of making an exact copy of a DNA molecule.

RNA (RIBONUCLEIC ACID): (Usually a single-stranded) molecule that transfers information carried by DNA to the cell's protein-making machinery.

RNA VACCINE: A new type of vaccine composed of the nucleic acid RNA packaged within a vehicle such as a lipid nanoparticle.

SEMISYNTHETIC ORGANISM: A genetically modified organism, often a bacterium like *E. coli*, and which has unnatural genetic information inserted into its genome.

SEQUENCING: To determine the primary structure of a polymer like DNA or RNA by analyzing the sequence of its four bases.

STEM CELL: An undifferentiated cell of an organism that is capable of giving rise to cells of many different tissues.

SYNTHETIC BIOLOGY: A discipline involved in the construction of new biological entities such as enzymes, genetic circuits, and cells or the redesign of existing biological systems. Synthetic biology builds on the advances in molecular, cell, and systems biology and seeks to transform biology in the same way that synthesis transformed chemistry and integrated circuit design transformed computing. The element that distinguishes synthetic biology is the focus on the design of core components that can be modeled and tuned to solve specific problems.

YEAST: Single-celled microorganisms classified as members of the fungus kingdom.

FURTHER READING

Atwood, Margaret. *Oryx and Crake*. New York: Anchor, 2004.

——. *The Year of the Flood*. London: Bloomsbury, 2009.

Baldwin, Geoff, Travis Bayer, Robert Dickinson, Tom Ellis, et al., eds. *Synthetic Biology: A Primer*. Singapore: World Scientific Publishing, 2012.

Baylis, Francoise. *Altered Inheritance: CRISPR and the Ethics of Human Genome Editing*. Cambridge, MA: Harvard University Press, 2019.

Carlson, Rob. *Biology Is Technology: The Promise, Peril, and the New Business of Engineering Life*. Cambridge, MA: Harvard University Press, 2010.

Chan, Alina, and Matt Ridley. *Viral: The Origin of COVID-19*. New York: HarperCollins, 2021.

Church, George, and Ed Regis. *Regenesis: How Synthetic Biology Will Reinvent Nature and Ourselves*. New York: Basic Books, 2012.

Contreras, Jorge. *The Genome Defense: The Epic Battle to Determine Who Owns Your DNA*. Chapel Hill, NC: Algonquin, 2019.

Cumbers, John and Karl Schmieder, *What's Your Biostrategy? How to Prepare Your Business For Synthetic Biology*, Emeryville, California: Pulp Bio Books, 2017

Davies, Jamie A. *Synthetic Biology: A Very Short Introduction*. Oxford: Oxford University Press, 2018.

——, ed. *Mammalian Synthetic Biology*. Oxford: Oxford University Press, 2019.

Doudna, Jennifer, and Samuel H. Sterne. *A Crack in Creation: Gene Editing and the Unthinkable Power to Control Evolution*. New York: Mariner, 2017.

Enriquez, Paul. *Rewriting Nature: The Future of Genome Editing and How to Bridge the Gap Between Law and Science*. Cambridge: Cambridge University Press, 2021.

Freemont, Paul, and Richard Kitney, eds. *Synthetic Biology: A Primer*. London: Imperial College Press, 2015.

Gibson, Daniel, Clyde Hutchison III, Hamilton Smith, and J. Craig Venter, eds. *Synthetic Biology: Tools for Engineering Biological Systems*. New York: Cold Spring Harbor Press, 2017.

Hockfield, Susan. *The Age of Living Machines: How Biology Will Build the Next Technology Revolution*. New York: Norton, 2019.

Hughes, Sally Smith. *Genentech: The Beginnings of Biotech*. Chicago: University of Chicago Press, 2011.

Isaacson, Walter. *Code Breaker: Jennifer Doudna, Gene Editing, and the Future of the Human Race*. New York: Simon and Schuster, 2021.

Keller, Evelyn Fox. *The Mirage of a Space Between Nature and Nurture*. Raleigh-Durham, NC: Duke University Press, 2010.

Kuldell, Natalie, Rachel Bernstein, Karen Ingram, and Kathryn Hart. *BioBuilder: Synthetic Biology in the Lab*. Sebastopol, CA: O'Reilly, 2015.

Markson, Sherri. *What Really Happened in Wuhan: A Virus Like No Other, Countless Infections, and Millions of Death*s. New York: HarperCollins, 2021.

Metzl, Jami. *Hacking Darwin: Genetic Engineering and the Future of Humanity*. Naperville, IL: Sourcebooks, 2019.

Rabinow, Paul. *Making PCR: A Story of Biotechnology*. Chicago: University of Chicago Press, 1997.

Redford, Kent, and William M. Adams. *Strange Natures: Conservation in the Era of Synthetic Biology*. New Haven, CT: Yale University Press, 2021.

Rutherford, Adam. *Creation: How Science Is Reinventing Life Itself*. New York: Current, 2013.

Roosth, Sophia. *Synthetic: How Life Got Made*. Chicago: University of Chicago Press, 2017.

Shapiro, Beth. *To Clone a Mammoth: The Science of De-Extinction*. Princeton, NJ: Princeton University Press, 2016.

Sheldrake, Merlin. *Entangled Life: How Fungi Make Our Worlds, Change Our Minds, and Shape Our Future*. New York: Random House, 2021.

Stamets, Paul. *Mycelium Running: How Mushrooms Can Help Save the World*. Berkeley, CA: Ten Speed Press, 2005.

Webb, Amy, and Andrew Hessel. *The Genesis Machine: Our Quest to Rewrite Life in the Age of Synthetic Biology*. New York: Public Affairs, 2021.

NOTES

INTRODUCTION

1. Danielle Tullman-Ercek, author interview, April 19, 2019.
2. Emily Hartman, author interview by phone, June 5, 2019.
3. Fred Kelly, *One Thing Leads to Another* (New York: Houghton Mifflin, 1936), 9; "Chaim Weizmann's Acetone Patent Turns 100," Weizmann Compass, September 27, 2015, https://www.weizmann.ac.il/Weizmann Compass/sections/people-behind-the-science/chaim-weizmann's -acetone-patent-turns-100.
4. E. Tambo, E. I. M. Khater, and J. H. Chen, "Nobel Prize for the Artemisinin and Ivermectin Discoveries: A Great Boost Towards Elimination of the Global Infectious Diseases of Poverty," *Infectious Diseases of Poverty* 4 (2015): 58, https://doi.org/10.1186/s40249-015-0091-8; V. Hale, J. D. Keasling, N. Renninger, et al., "Microbially Derived Artemisinin: A Biotechnology Solution to the Global Problem of Access to Affordable Antimalarial Drugs," in *Defining and Defeating the Intolerable Burden of Malaria III: Progress and Perspectives: Supplement to Volume 77(6) of American Journal of Tropical Medicine and Hygiene*, ed. J. G. Breman, M. S. Alilio, and N. J. White (Northbrook, IL: American Society of Tropical Medicine and Hygiene, 2007), https://www.ncbi .nlm.nih.gov/books/NBK1717/.
5. Andrew Ward, quoted in William Booth and Carolyn Johnson, "Elegant but Unproven: RNA Experiments Leap to the Front in

Coronavirus Vaccine Race," *Washington Post*, July 5, 2020, https://www.washingtonpost.com/world/europe/coronavirus-vaccine-race-messenger-rna-imperial-college/2020/07/05/6565b2d0-ba2e-11ea-97c1-6cf116ffe26c_story.html.

6. White House, "President Biden to Launch a National Biotechnology and Biomanufacturing Initiative," September12, 2022, https://www.whitehouse.gov/briefing-room/statements-releases/2022/09/12/fact-sheet-president-biden-to-launch-a-national-biotechnology-and-biomanufacturing-initiative/.

1. A GLASS OF ABSINTHE: A MALARIA MEDICINE

1. Jay Keasling, author interview via WebEx, April 24, 2019.

2. Eric Hobsbawm, *The Industrial Revolution* (Oxford: Oxford University Press, 1971), 39.

3. Daniel Grushkin, "The Rise and Fall of the Company That Was Going to Have Us All Using Biofuels," *Fast Company*, August 8, 2012, https://www.fastcompany.com/3000040/rise-and-fall-company-was-going-have-us-all-using-biofuels.

4. Evelyn Fox Keller, *Making Sense of Life* (Cambridge, MA: Harvard University Press, 2002), 24.

5. Evelyn Fox Keller, "Making as Knowing; Knowing as Making," *Biological Theory* 4, no. 4 (2009): 333–39.

6. Tom Volk, "Tom Volk's Fungus of the Month for November, 2003," https://botit.botany.wisc.edu/toms_fungi/nov2003.html.

7. Sally Bedell Smith, *Genentech: The Beginnings of Biotech* (Chicago: University of Chicago Press, 2010), 44.

8. "A Growing Number of Governments Hope to Clone America's DARPA," *The Economist*, June 3, 2021, https://www.economist.com/science-and-technology/2021/06/03/a-growing-number-of-governments-hope-to-clone-americas-darpa.

9. "IDR Team Summary 4," in *Synthetic Biology: Building on Nature's Inspiration* (Washington, DC: National Academies Press, 2010), 42, https://haseloff.plantsci.cam.ac.uk/resources/SynBio_reports/Naturesinspiration.pdf.

10. Michael Spector, "A Life of Its Own," *New Yorker*, September 28, 2009, https://www.newyorker.com/magazine/2009/09/28/a-life-of-its-own.

11. Wendell Lim, author interview via Zoom, December 6, 2021.

12. "Oil Price 'May Hit $200 a Barrel," BBC News, last updated May 7, 2008, http://news.bbc.co.uk/2/hi/business/7387203.stm.

13. Jay Keasling, quoted in *Nova ScienceNOW*, February 23, 2011.

14. Jay Keasling, author interview, August 5, 2019.

15. UNICEF, "Malaria," https://data.unicef.org/topic/child-health/malaria/, last updated July 2022.

16. D. K. Ro, E. M. Paradise, M. Ouellet, et al., "Production of the Anti-Malarial Drug Precursor Artemisinic Acid in Engineered Yeast," *Nature* 440 (April 2006): 940–43, https://doi.org/10.1038/nature04640.

17. C. Paddon, P. Westfall, D. Pitera, et al., "High-Level Semi-Synthetic Production of the Potent Antimalarial Artemisinin," *Nature* 496 (April 2013): 528–32, https://doi.org/10.1038/nature12051.

18. Adam Arkin, author interview via Zoom, July 2, 2021.

19. Neil Renninger, quoted in Grushkin, "The Rise and Fall."

20. John Melo, quoted in Grushkin, "The Rise and Fall."

21. Jim Lane, "Amyris Inks 5-Year $100M+ Biofene Supply Pact for Nutraceutical Market," *Biofuels Digest*, April 28, 2016, https://www.biofuelsdigest.com/bdigest/2016/04/28/amyris-inks-5-year-100m-biofene-supply-pact-for-nutraceutical-market/.

22. Fabiano Piccinno, Roland Hischier, Stefan Seeger, and Claudia Som, "From Laboratory to Industrial Scale: A Scale-Up Framework for Chemical Processes in Life Cycle Assessment Studies," *Journal of Cleaner Production* (2016), https://www.researchgate.net/publication/304705424 _From_laboratory_to_industrial_scale_a_scale-up_framework_for _chemical_processes_in_life_cycle_assessment_studies.

23. Grushkin, "The Rise and Fall."

24. Timothy Gardner, author interview via Zoom, December 1, 2021.

25. Author visit, Joint BioEnergy Institute, Emeryville, California, August 1, 2019.

26. Quentin Hardy, "Clean Tech Rises Again, Retooling Nature for Industrial Use," *New York Times*, December 27, 2016, https://www.nytimes.com/2016/12/27/technology/clean-tech-rises-again-retooling-nature-for-industrial-use.html.

27. Newmark West (@Newmark_West), "#Zymergen Grows to 800 Employees; Moves to 303,000 SF Former Chiron Space in Emeryville. NKF Brokers Bill Benton, Mike Brown, Jennifer Vergara & Mary

Hines in the News," Twitter, October 15, 2019, https://twitter.com
/newmark_west/status/1184202769026166789.

28. Jeffrey Dietrich, "Malonic Acid: A Bioadvantaged Chemical," Lygos,
December 2014, https://lygos.com/app/uploads/2015/03/2014-12-Website
-Flyer.pdf

29. Eric Steen, author interview, August 2, 2019.

30. Lygos, Press Release, "Lygos Awarded $300,000 Small Business Voucher
from DOE," March 11, 2016, https://lygos.com/265-2.

31. Cision, "Squalene Market to Grow by Over $55 Million: COVID
Impact Analysis," Cision PR Newswire, April 6, 2021, https://www
.prnewswire.com/news-releases/squalene-market-to-grow-by-over
--55-million--covid-19-impact-analysis-key-drivers-trends--major
-vendor-offerings-technavio-301261829.html.

32. Jay Keasling, author interview, April 15, 2022.

2. A RADICAL PHILOSOPHY

1. Pamela Silver, author interview by phone, April 11, 2019.

2. Lisa Robinson, "An Oral History of Laurel Canyon, the 60s and 70s
Music Mecca," *Vanity Fair*, March 2015, https://www.vanityfair.com
/culture/2015/02/laurel-canyon-music-scene; Joel Selvin, "The Top 100
Bay Area Bands / The '70s," SFGATE.com, https://www.sfgate.com
/bayarea/article/The-Top-100-Bay-Area-Bands-The-70s-3312126.php.

3. Pamela Silver, author interview by phone, April 11, 2019.

4. Margaret Nelson and Pamela Silver, "Context Affects Nuclear Protein
Localization of *Saccharomyces cerevisiae*," *Molecular and Cellular Biology* 9,
no. 2 (February 1989): 364–89, https://www.ncbi.nlm.nih.gov/pmc
/articles/PMC362612/pdf/molcellb00050-0038.pdf.

5. Pamela Silver, quoted in Christina Agapakis, "10 Tips to Transform
Your Career and Science Culture," *NeoLife*, July 8, 2021, https://neo
.life/2021/07/10-tips-to-transform-your-career-and-science-culture
/https://neo.life/2021/07/10-tips-to-transform-your-career-and
-science-culture/.

6. L. Chan, S. Kosuri, and D. Endy, "Refactoring T7 Bacteriophage T7,"
Molecular Systems Biology, (2005) 1:2005-0018, Sept 13, 2005, https://
www.ncbi.nlm.nih.gov/pmc/articles/PMC1681472/.

7. Michael Elowitz, author interview via Zoom, October 26, 2020.

8. Timothy Gardner, author interview via Zoom, December 1, 2021.

9. Elowitz, author interview.

10. Gardner, author interview.

11. T. Gardner, C. Cantor, and J. Collins, "Construction of a Genetic Toggle Switch in *Escherichia coli*," *Nature* 403 (January 2000): 339–42, https://doi.org/10.1038/35002131; M. Elowitz and S. Liebler, "A Synthetic Oscillatory Network of Transcriptional Regulators," *Nature* 403, (January 2000): 335–38, https://www.nature.com/articles/35002125.

12. Gardner, Cantor, and Collins, "Construction of a Genetic Toggle Switch," 339–42.

13. Mo Khalil, author interview via Zoom, July 18, 2022.

14. J. Y. Chan, S. Kosuri, and D. Endy, "Refactoring Bacteriophage T7," *Molecular Systems Biology* (September 13, 2005): 2005.0018, https://www.ncbi.nlm.nih.gov/pmc/articles/PMC1681472/pdf/msb4100025.pdf.

15. Drew Endy, quoted in Sophia Roosth, *Synthetic: How Life Got Made* (Chicago: University of Chicago Press, 2015), 24.

16. David Ewing Duncan, "Is the World Ready for Synthetic People?" *NeoLife*, April 5, 2018, https://neo.life/2018/04/is-the-world-ready-for-synthetic-people/.

17. Tom Knight, Panel discussion at Ferment 2021 Conference, Boston, Massachusetts, October 28, 2021.

18. Pamela Silver, author interview

19. Will Wright, iGEM director of entrepreneurship, "Built With Biology" presentation, April 13, 2022.

20. iGEM, "Congratulations to All iGEM 2020 Participants!" https://2020.iGEM.org/Main_Page.

21. iGEM, "Teams: Welcome to the iGEM 2021 Website," https://2021.iGEM.org/Main_Page.

22. Amy Feldman, "The Life Factory: Synthetic Organisms From \$1.4 Billion Startup Will Revolutionize Manufacturing," *Forbes*, August 5, 2019, https://www.forbes.com/sites/amyfeldman/2019/08/05/the-life-factory-synthetic-organisms-from-startup-ginkgo-bioworks-unicorn-will-revolutionize-manufacturing/?sh=291f9d3d145e.

23. Jason Kelly, conversation with author at FERMENT Conference, October 28, 2021.

24. Reshma Shetty, author interview by phone, November 15, 2019.

25. Reshma Shetty, author interview by phone, November 15, 2019.

26. Reshma Shetty, quoted in "The Future of Biotechnology: Solutions for Energy, Agriculture and Manufacturing," hearing before the House Committee on Science, Space, and Technology, 114th Congress, 32 (December 8, 2015) [testimony of Reshma Shetty, cofounder of Ginkgo Bioworks], https://www.govinfo.gov/content/pkg/CHRG-114hhrg20824 /html/CHRG-114hhrg20824.htm.

27. Author visit, Ferment Conference, Boston, MA, October 13, 2019.

28. Daniel Nocera, author interview via Zoom, July 8, 2022.

29. Dan Nocera, quoted in A. Jia and S. Narayanan, "Harvard Researchers Pioneer Photosynthetic Bionic Leaf," *Harvard Crimson*, February 16, 2018, https://www.thecrimson.com/article/2018/2/16/bionic-leaf/.

30. Laurie Zoloth, interviewed by Katheryn Brink and Adam Silverman, "In Translation," Engineering Biology Research Consortium Podcast, May 3, 2021, https://podcasts.apple.com/us/podcast/4-bioethics-and -flesh-eating-worms-w-laurie-zoloth/id1552148593?i=1000519968049.

31. Drew Endy, "Synthetic Biology: What Should We Be Vibrating About?" TEDx Stanford, June 2014 [video], 8:53, https://engineering.stanford .edu/magazine/article/drew-endy-discusses-what-bioengineers -should-be-vibrating-about.

32. Integrated DNA Technologies, "About Us," https://www.idtdna.com /pages/about (accessed October 31, 2022).

33. Sam Kass, quoted in Luke Timmerman, "From Antibiotics to Food Safety, Sample6 Finds A Niche," *Forbes*, August 22. 2016, www .forbes.com/sites/luketimmerman/2016/08/22/from-antibiotics -to-food-safety-sample6-finds-a-niche-and-12-7m-after-a-sharp -turn/?sh=138a16623509.

34. Sam Kass, quoted in "Sample6 Announces Dr. Michael Koeris as CEO," Talent4Boards, August 22, 2016, https://talent4boards.com/sample6 -announces-dr-michael-koeris-ceo-sam-kass-board-member-along -closing-12-5m-series-c-financing-led-acre-venture-partners/.

3. PANDORA'S BOX: THE TRIUMPH AND TEMPTATION OF GENE EDITING

1. Rob Stein, "A Young Mississippi Woman's Journey Through a Gene Editing Therapy," *All Things Considered*, NPR, December 25, 2019 [audio],

https://www.npr.org/sections/health-shots/2019/12/25/784395525/a
-young-mississippi-womans-journey-through-a-pioneering-gene
-editing-experiment.

2. Stein, "A Young Mississippi Woman's Journey."

3. Eva Nevelius, "Press Release: The Nobel Prize in Chemistry 2020," Nobelprize.org, October 7, 2020, https://www.nobelprize.org/prizes/chemistry/2020/press-release/.

4. H. Frangoul, D. Altshuler, M.D. Cappellini, Y.S. Chen, "CRISPR-Cas9 Gene Editing for Sickle Cell Disease and b-Thalassemia," *New England Journal of Medicine*, no. 384 (January 21, 2021): 252–60, https://www.nejm.org/doi/full/10.1056/NEJMoa2031054.

5. Alex Phillippidis, "Making History with the 1990 Gene Therapy Trial," *Genetic Engineering News*, March 10, 2016, https://www.genengnews.com/magazine/269/making-history-with-the-1990-gene-therapy-trial/.

6. Emma Young, "Miracle Gene Therapy Trial Halted," *New Scientist*, October 30, 2002, https://www.newscientist.com/article/dn2878-miracle-gene-therapy-trial-halted/.

7. Meir Rinde, "The Death of Jesse Gelsinger, 20 Years Later," Science History Institute, June 4, 2019, https://www.sciencehistory.org/distillations/the-death-of-jesse-gelsinger-20-years-later.

8. Jennifer Doudna, quoted in Walter Isaacson, *The Codebreaker: Jennifer Doudna, Gene Editing, and the Future of the Human Race* (New York: Simon and Schuster, 2021), 105.

9. Jennifer Doudna, *A Crack in Creation: Gene Editing and the Unthinkable Power to Control Evolution* (New York: HarperCollins, 2017), 22.

10. Emmanuelle Charpentier, "Emmanuelle Charpentier Founder of the CRISPR-Cas9 Gene Editing Technology," by Dramasvecia for the Knut and Alice Wallenberg Foundation, YouTube, October 7, 2020 [video], 0:55–1:09, https://youtu.be/7NCniou32Q4.

11. Charpentier, "Founder of the CRISPR-Cas9 Gene Editing Technology."

12. Alison Abbott, "The Quiet Revolutionary: How the Co-discovery of CRISPR Explosively Changed Emmanuelle Charpentier's Life," *Nature* 532 (2016): 432–34, https://www.nature.com/articles/532432a.

13. M. Jinek, E. Charpentier, J. Doudna, et al., "A Programmable Dual-RNA-Guided DNA Endonuclease in Adaptive Bacterial Immunity,"

Science 337, no. 6096 (August, 2012): 816–21, https://www.science.org /doi/10.1126/science.1225829.

14. Hui Yang, Haoyi Wang, Chikdu S. Shivalila, et al., "One-Step Generation of Mice Carrying Reporter and Conditional Alleles by CRISPR/ Cas-Mediated Genome Engineering," *Cell* 154, no. 6 (2013): P1370–79, https://www.cell.com/fulltext/S0092-8674(13)01016-7.

15. Jon Cohen, "How the Battle Lines Over CRISPR Were Drawn," *Science*, February 12, 2017, https://www.science.org/content/article/how -battle-lines-over-crispr-were-drawn.

16. Drew Endy and Laurie Zoloth, "Should We Synthesize a Human Genome?" *Cosmos*, May 12, 2016, https://cosmosmagazine.com/people /society/should-we-synthesise-a-human-genome/.

17. Henry T. Greedy, "Of Science, CRISPR-Cas9, and Asilomar," Stanford Law School Blog, April 4, 2015, https://law.stanford.edu/2015/04/04 /of-science-crispr-cas9-and-asilomar/.

18. Ben Hurlbut, presentation at Northwestern University Mammalian Synthetic Biology Conference, Evanston, Illinois, June 4, 2019.

19. Isaacson, *Codebreaker*, 304–6.

20. Jon Cohen, "The Untold Story of the 'Circle of Trust' Behind the World's First Gene-Edited Babies," *Science*, August 1, 2019, https:// www.science.org/content/article/untold-story-circle-trust-behind -world-s-first-gene-edited-babies?adobe_mc=MCMID%3D01366832 0860573977892095321058999621172%7CMCORGID%3D242B647254119 9F70A4C98A6%2540AdobeOrg%7CTS%3D1669744906.

21. Antonio Regalado, "The Creator of the CRISPR Babies Has Been Released from Prison," *MIT Technology Review*, April 4, 2022, https:// www.technologyreview.com/2022/04/04/1048829/he-jiankui-prison -free-crispr-babies/.

22. Leeor Kaufman, dir., *Unnatural Selection*, Radley Studios, 2019, https:// www.netflix.com/watch/80208833?source=35&trackId=254743534.

23. Victoria Gray, quoted in Amy Dockser Marcus, "CRISPR's Next Frontier. Treating Common Conditions," *Wall Street Journal*, May 5, 2021, https://www.wsj.com/articles/crisprs-next-frontier-treating-common -conditions-11620226832; Stein, "A Young Mississippi Woman's Journey"; H. Frangoul, D. Altshuler, M. D. Cappellini, et al., "CRISPR Cas9 Gene Editing for Sickle Cell Disease and β-Thalassemia," 384, no. 3

New England Journal of Medicine (December 2020): 252–60, https://www.nejm.org/doi/pdf/10.1056/NEJMoa2031054.

24. Victoria Gray and Haydar Frangoul quoted in Roz Plater, "First Person Treated for Sickle Cell Disease With CRISPR Is Doing Well," *Healthline*, July 6, 2020, https://www.healthline.com/health-news/first-person-treated-for-sickle-cell-disease-with-crispr-is-doing-well.

25. Biotechnology Information Organization, "Synthetic Biology Explained," https://archive.bio.org/articles/synthetic-biology-explained (accessed November 1, 2022).

26. Annie Melchior, "WHO Releases New Guidelines on Human Genome Editing," *The Scientist*, July 12, 2021, https://www.the-scientist.com/news-opinion/who-releases-new-recommendations-on-human-genome-editing-68964.

27. M. Stefanidakis, M. Maeder, George Bounoutas, et al., "Preclinical Assessment of *In Vivo* Gene Editing Efficiency, Specificity, and Tolerability of EDIT-101, an Investigational CRISPR Treatment for Leber Congenital Amaurosis 10 (LCA10)," Editas Medicine, https://www.editasmedicine.com/wp-content/uploads/2019/10/16.pdf (accessed November 1, 2022).

28. Marcus, "CRISPR's Next Frontier."

29. Jon Cohen, "The Latest Round in the CRISPR Patent Battle Has an Apparent Victor, but the Fight Continues," *Science*, September 11, 2020, https://www.sciencemag.org/news/2020/09/latest-round-crispr-patent-battle-has-apparent-victor-fight-continues.

30. Crisprtx, Press Release, "Vertex and CRISPR Therapeutics Establish Collaboration to Use CRISP-Cas9 Gene Editing Technology to Discover and Develop New Treatments for Genetic Diseases," October 26, 2015 https://crisprtx.com/about-us/press-releases-and-presentations/vertex-and-crispr-therapeutics-establish-collaboration-to-use-crispr-cas9-gene-editing-technology-to-discover-and-develop-new-treatments-for-genetic-diseases-1.

31. B. E. Rubin, S Diamond, B. F. Cress, et al., "Species- and Site-Specific Genome Editing in Complex Bacterial Communities," *Nature Microbiology* 7 (2022): 34–47, https://doi.org/10.1038/s41564-021-01014-7.

32. K. Li, A. Scarano, N. M. Gonzalez, et al., "Biofortified Tomatoes Provide a New Route to Vitamin D Sufficiency," *Nature Plants* 8 (2022): 611–616. https://doi.org/10.1038/s41477-022-01154-6

33. Rob Stein, "A Year in, 1st Patient to Get Gene Editing for Sickle Cell Disease Is Thriving," *Morning Edition*, NPR, June 23, 2020 [audio], https://www.npr.org/sections/health-shots/2020/06/23/877543610/a-year-in-1st-patient-to-get-gene-editing-for-sickle-cell-disease-is-thriving.

4. THE SILK ROAD: DIRECTING EVOLUTION

1. Christopher Voigt, "Synthetic Biology: Scale-Up and Defense Applications," YouTube, January 22, 2016 [video], https://m.youtube.com/watch?v=LCwS3yFkSys.

2. Christopher Voigt, author interview by phone, July 29, 2022

3. Frances Arnold, "The Nobel Prize: Frances H. Arnold," Nobelprize.org, 2018, https://www.nobelprize.org/womenwhochangedscience/stories/frances-arnold.

4. Voigt Lab, http://web.mit.edu/voigtlab/ (accessed November 1, 2022).

5. Arnold, "The Nobel Prize: Frances H. Arnold."

6. Arnold, "The Nobel Prize: Frances H. Arnold."

7. Diana Kormos-Buchwald, quoted in Natalie Angier, "Frances Arnold Turns Microbes into Living Factories," *New York Times*, May 28, 2019, https://www.nytimes.com/2019/05/28/science/frances-arnold-caltech-evolution.html.

8. Frances Arnold, quoted in Natalie Angier, "Frances Arnold Turns Microbes into Living Factories," *New York Times*, May 28, 2019, https://www.nytimes.com/2019/05/28/science/frances-arnold-caltech-evolution.html.

9. Christopher Voigt, author interview

10. David Breslauer, author interview by phone, July 17, 2019.

11. Colin Emery, "Silkworm Thread vs. Spider Thread: Strength and Future Uses," Everything Silkworms.com, September 13, 2019, https://everythingsilkworms.com.au/silkworm-thread-vs-spider-thread/.

12. David Breslauer, author interview via Zoom, June 15, 2021.

13. Brian McEvoy, "Arachnid Ales Uses Yeast to Make Spider Silk," Hackaday.com, November 5, 2018, https://hackaday.com/2018/11/05/arachnid-ale-uses-yeast-to-make-spider-silk/.

14. Christopher Voigt, quoted in Amy Feldman, "Clothes from a Petri Dish: $700 Million Bolt Threads May Have Cracked the Code on Spider

Silk, *Forbes*, August 15, 2018, https://www.forbes.com/sites/amyfeldman /2018/08/14/clothes-from-a-petri-dish-700-million-bolt-threads-may -have-cracked-the-code-on-spider-silk/?sh=52e93e00bda1.

15. Sean Simpson, author interview, November 10, 2021.

16. D. Jones and S. Kreis, "Origins and Relationships of Industrial Solvent-Producing Clostridial Strains," *FEMS Microbiology Reviews* 17, no. 3 (October 1995): 223–32, https://doi.org/10.1111/j.1574-6976.1995.tb00206.x.

17. David Breslauer, author interview by phone, July 17, 2019.

18. D. M. Widmaier, D. Tullman-Ercek, E. A. Mirsky, et al., "Engineering the Salmonella Type III Secretion System to Export Spider Silk Monomers," *Molecular Systems Biology* 5 (September 2009): article 309, https://www.embopress.org/doi/epdf/10.1038/msb.2009.62.

19. Danielle Tullman-Ercek, author interview via Zoom, July 21, 2021.

20. David Breslauer, author interview via Zoom, June 15, 2021.

21. Lisa M. Krieger, "UC Berkley Loses CRISPR Gene Editing Patent Case," *San Jose Mercury News*, February 28, 2022, https://www.mercurynews .com/2022/02/28/uc-berkeley-loses-crispr-gene-editing-patent-case/amp/.

22. Zachary Serber, author interview, August 2, 2019.

23. Nancy Kelley-Loughnane, "Future of Material Synthesis for Military Environments," Presentation at SEED Conference 2019, New York, NY, June 25, 2019.

24. Kelley-Loughnane, "Future of Material Synthesis for Military Environments,"

25. Christina Agapakis, "Love Our Monsters—Radical Collaboration in a Post-Disciplinary Age," Synthetic Biology Leadership Excellence Accelerator Program, Version 1, January 28, 2013, https://static1.squarespace .com/static/58b1ecdeebbd1abb370019db/t/593981dc3e00be2ed1de1 63d/1496941021049/love-our-monsters-radical-collaboration-in-a -post-disciplinary-age.pdf

26. Jason Kelly, "I Run a GMO Company—And I Support GMO Labeling," *New York Times*, May 16, 2016, A32, https://www.nytimes .com/2016/05/16/opinion/i-run-a-gmo-company-and-i-support-gmo -labeling.html.

27. A. A. K. Nielsen, B. S. Der, J. Shin, et al., "Genetic Circuit Design Automation," *Science* 352, no. 6281 (April 2016): aac7341, https://www .academia.edu/58410617/Genetic_circuit_design_automation?auto =citations&from=cover_page.

28. Michael Köpke, author visit to LanzaTech, Skokie, IL, July 30, 2019.

29. Bethany Halford, "Frances Arnold on Founding a Company: Building the Right Team Is Critical," *Chemical and Engineering News*, March 8, 2020, https://cen.acs.org/biological-chemistry/synthetic-biology/Frances -Arnold-on-founding-a-company-Building-the-right-teams-is -critical/98/i9.

30. Ahmad (Mo) Khalil, interview by author via Zoom, July 18, 2022.

31. "Beef Fat to Aviation Fuel," *Iowa Farmer Today*, May 10, 2022, https://www .agupdate.com/iowafarmertoday/news/livestock/beef-fat-converted-to -aviation-fuel/article_cb0282e8-ccac-11ec-aca8-d7a3c525fdb4.html.

5. WILD: REMAKING LIFE

1. Author visit, Twist Bioscience, South San Francisco, CA, June 15, 2022.

2. F. H. Crick, "The Origin of the Genetic Code," *Journal of Molecular Biology* 38 (1968): 367–79, doi: 10.1016/0022-2836(68)90392-6.

3. "Understanding the Rules of Life: Building a Synthetic Cell," Program Solicitation NSF 18–599, National Science Foundation, https://www .nsf.gov/pubs/2018/nsf18599/nsf18599.htm (accessed October 22, 2021).

4. J. Craig Venter, quoted in Ed Yong, "The Mysterious Thing About a Marvelous New Synthetic Cell," *The Atlantic*, March 2016, https:// www.theatlantic.com/science/archive/2016/03/the-quest-to-24make -synthetic-cells-shows-how-little-we-know-about-life/475053/.

5. "Ugur Sahin and Özlem Türeci: Meet the Scientist Couple Driving an mRNA Vaccine Revolution," Chris Anderson, TED, YouTube [video], https://www.youtube.com/watch?v=VdqnAhNrqPU (accessed November 2, 2022).

6. Steven Benner, author interview via Zoom, August 11, 2021.

7. John Glass, author interview via Zoom, July 20, 2021.

8. D. G. Gibson, J. I. Glass, C. Lartigue, et.al. "Creation of a Bacterial Cell Controlled by a Chemically Synthesized Genome," *Science* 329, no. 5987 (July 2, 2010): 52–56, https://science.sciencemag.org/content /sci/329/5987/52.full.pdf.

9. "Genesis Redux," *The Economist*, May 20, 2010, 81–83, https://www .economist.com/briefing/2010/05/20/genesis-redux.

10. Craig Venter, *Life at the Speed of Light: From the Double Helix to the Dawn of Life* (New York: Crown Books, 2017), 67.

11. C. A. Hutchison III, R-Y. Chuang, V. N. Noskov, et.al., "Design and Synthesis of a Minimal Bacterial Genome," *Science* 351, no. 6280 (March 25, 2016), https://science.sciencemag.org/content/351/6280/aad6253.

12. E. Martinez-Garcia, V. de Lorenzo, "The Quest For the Minimal Bacterial Genome," *Current Opinion in Biotechnology*, 42: 216–24, https://www.sciencedirect.com/science/article/abs/pii/S0958166916301859?via%3Dihub.

13. "NASA-Funded Research Creates DNA-like Molecule to Aid Search for Alien Life," NASA News, February 22, 2019, https://solarsystem.nasa.gov/news/859/nasa-funded-research-creates-dna-like-molecule-to-aid-search-for-alien-life/.

14. Michael Jewett, quoted in Ruth Williams, "DNA's Coding Power Doubled," *The Scientist*, February 21, 2019, https://www.the-scientist.com/news-opinion/dnas-coding-power-doubled-65499,

15. Katalin Karikó, quoted in Gina Kolata, "Kati Karikó Helped Shield the World from Coronavirus," *New York Times*, April 9, 2021, https://www.nytimes.com/2021/04/08/health/coronavirus-mrna-kariko.htm.

16. Julia Kollewe, "Covid Vaccine Technology Pioneer: I Never Doubted It Would Work," The Guardian, November 21, 2020, https://www.theguardian.com/science/2020/nov/21/covid-vaccine-technology-pioneer-i-never-doubted-it-would-work.

17. K. Karikó, M. Buckstein, H. Ni, et al., "Suppression of RNA Recognition by Toll-like Receptors: The Impact of Nucleoside Modification and the Evolutionary Origin of RNA," *Immunity* 23 (August 2005): 165–75, https://www.sciencedirect.com/science/article/pii/S1074761305002116.

18. Elie Dolgin, "The Tangled History of mRNA," *Nature* 597 (September 14, 2021): 318–24, https://www.nature.com/articles/d41586-021-02483-w.

19. Dolgin, "Tangled History of mRNA."

20. Joe Miller, "Inside the Race for A COVID Vaccine: How BioNTech Made a Breakthrough," *Financial Times*, November 13, 2020, https://www.ft.com/content/c4ca8496-a215-44b1-a7eb-f88568fc9de9.

21. Ugur Sahin, quoted in David Gelles, "If the Pandemic Had Hit a Year Earlier, We Might Not Have Been in a Position to Respond This Fast. Here's Why," *Atlantic Council*, November 8, 2021, https://www.atlanticcouncil.org/blogs/new-atlanticist/if-the-pandemic-hit-a-year-earlier-we-might-not-have-been-in-the-position-to-respond-this-fast-say-biontech-co-founders-heres-why/.

22. Joanna Roberts, "Q&A: BioNTech Vaccine Is Only 'mRNA 1.0.' This Is Just the Beginning, Say Co-Founders," *Horizon: The EU Research and Innovation Magazine*, April 8, 2021, https://ec.europa.eu/research -and-innovation/en/horizon-magazine/qa-biontech-vaccine-only -mrna-10-just-beginning-say-co-founders.

23. "The BioNTech Approach: How We Enable Our Vision," BioNTech, https://biontech.de/how-we-translate/biontech-approach (accessed October 22, 2021).

24. Floyd Romesberg, author interview by phone, December 12, 2019.

25. Emily Leproust, "Q & A with Fiona Mischel," Built with Biology Global Conference, Oakland, CA, April 14, 2022.

26. Leproust, "Q & A with Fiona Mischel."

27. Angela Bitting and James Diggans, author interview via Zoom, July 2, 2021.

28. "TWIST—Twist Bioscience Corporation," Yahoo Finance, https:// finance.yahoo.com/quote/TWST/key-statistics/.

29. "Deep Purple's 'Smoke on the Water' Becomes a Piece of Scientific History," Twist Bioscience, December 12, 2017, https://www.twistbioscience .com/blog/company-news-updates/deep-purples-smoke-water -becomes-piece-scientific-history.

30. Zachary Serber, author interview, August 2, 2019.

31. Damian Garde and Jonathan Saltzman, "The Story of mRNA: How a Once-Dismissed Idea Became a Leading Technology in the Covid Vaccine Race," STAT News, November 10, 2020, https://www.statnews .com/2020/11/10/the-story-of-mrna-how-a-once-dismissed-idea -became-a-leading-technology-in-the-covid-vaccine-race/.

32. Jason Schrum, quoted in Kelly Servick, "This Mysterious $2 Billion Biotech Is Revealing the Secrets Behind Its New Drugs and Vaccines," *Science*, March 25, 2020, https://www.science.org/content/article/mysterious -2-billion-biotech-revealing-secrets-behind-its-new-drugs-and -vaccines.

33. John Glass, e-mail to author, September 2, 2022.

34. J. Craig Venter, John I. Glass, Clyde A. Hutchison III, et al., "Synthetic Chromosomes, Genomes, Viruses, and Cells," *Cell* 185, no. 15 (2022): 2708–24.

6. RUSH: BIOLOGY-MADE MEDICINES

1. Stephanie Culler, author interview via Zoom, February 10, 2021.

2. Culler, author interview.

3. Culler, author interview

4. S. Galanie, K. Thodey, I. Trenchard, *et al.*, "Complete Biosynthesis of Opioids in Yeast," *Science* 349, no. 6252 (2015): 1095–1100, https://www.ncbi.nlm.nih.gov/pmc/articles/PMC4924617/.

5. "CHAIN to Explore Manufacturing of Novel Medicines," CHAIN Biotechnology, March 26, 2018, https://chainbiotech.com/2018/03/26/chain-to-explore-manufacturing-of-novel-medicines/.

6. Zachary Sun, presentation at SEED Conference, New York, NY, June 11, 2019.

7. Culler, author interview.

8. Jacelyn Walsh, quoted in Jamie Bartosch, "After Years of Treatment and Three Relapses, CAR-T-Cell Therapy Offers Hope for Leukemia Patient," UChicago Medicine, October 23, 2019, https://www.uchicagomedicine.org/forefront/cancer-articles/CAR-T-cell-therapy-offers-hope-for-leukemia-patient.

9. "CAR T Cell Therapy, " Penn Medicine, https://www.pennmedicine.org/cancer/navigating-cancer-care/treatment-types/immunotherapy/what-is-car-t-therapy (accessed November 3, 2022).

10. "AbbVie and Caribou Enter $340m Deal to Develop CAR-T Cell Therapies," *Pharmaceutical Technology*, February 11, 2021, https://www.pharmaceutical-technology.com/news/abbvie-caribou-340m-deal/.

11. "Kymriah," Novartis, https://www.hcp.novartis.com/products/kymriah/acute-lymphoblastic-leukemia-children/.

12. "Immunotherapy for Cancer: CAR-T-Cell Therapy," OHSU Knight Cancer Institute, https://www.ohsu.edu/knight-cancer-institute/CAR-T-cell-therapy-cancer (accessed November 3, 2022).

13. L. Morsut, K. Roybal, W. Lim, et al., "Engineering Customized Cell Sensing and Response Behaviors Using Synthetic Notch Receptors," *Cell* 164 (February 11, 2016): 780–91, https://www.cell.com/cell/pdfExtended/S0092-8674(16)00052-0.

14. Leonardo Morsut, author interview via Zoom, March 15, 2021.

15. Wendell Lim, author interview via Zoom. December 6, 2021.

16. Arie Belldegrun, presentation at Ginkgo FERMENT Conference, Boston, MA, October 28, 2021.

17. Antonio Regalado, "How Synthetic Biology Went From No Way to Pay Day in the Cannabis Business," *MIT Technology Review*, May 13, 2019, https://www.technologyreview.com/2019/05/13/975/how-biotech -went-from-no-way-to-payday-in-the-cannabis-business/.

18. Christina Smolke, "Finding Medicine Where You Least Expect It," presentation at TEDxStanford, YouTube [video], https://www.youtube .com/watch?v=gIuhIo7mO5A (accessed November 3, 2022).

19. Kristy Hawkins, author interview, April 7, 2022.

20. Mike Gorenstein, presentation at Ginkgo FERMENT Conference, Boston, MA, October 28, 2021.

21. "Demetrix Secures $50 Million in Series A Funding," Demetrix, July 11, 2019, https://demetrix.com/demetrix-secures-50-million-in-series -a-funding-to-provide-cannabinoids/.

22. Declan Bates, quoted in Luke Walton, "Drug-Producing Bacteria Possible with Synthetic Biology Breakthrough" Phys.org, March 5, 2018, https://phys.org/news/2018-03-drug-producing-bacteria-synthetic -biology-breakthrough.html.

23. The quotations from Deans and Seamons that follow are from my visit to the Tara Deans lab at the University of Utah, July 13, 2021.

24. "Consumer Alert on Regenerative Medicine Products Including Stem Cells and Exomes," FDA, last modified July 22, 2020, https:// www.fda.gov/vaccines-blood-biologics/consumers-biologics /consumer-alert-regenerative-medicine-products-including-stem -cells-and-exosomes.

25. Jay Gopalakrishnan, quoted in Michelle Starr, "Scientists Grew Stem Cell Mini-Brains. Then the Brains Sort-of Developed Eyes," Science Alert, August 17, 2021, https://www.sciencealert.com/scientists-used -stem-cells-to-make-mini-brains-they-grew-rudimentary-eyes.

26. Aoife Brennan, author interview by phone, September 1, 2020.

27. Synlogic, news release, "Synlogic Announces SYNB1353 Achieves Proof of Mechanism For Treatment of Homocystinuria and Provides Business Update," November 30, 2022, https://investor.synlogictx.com/news -releases/news-release-details/synlogic-announces-synb1353-achieves -proof-mechanism-treatment.

28. Culler, author interview.

29. Sarah Elizabeth Richards, "How the Microbiome Could Be the Key to New Cancer Treatment," *Smithsonian*, March 7, 2019, https://www.smithsonianmag.com/science-nature/how-microbiome-could-be-key-new-cancer-treatments-180971589/.

30. Kristala Prather, author interview by phone, February 10, 2021.

31. Dawn Ericksen, Sijin Li, and Huimin Zhao, "Chapter 3: Pathway Engineering as an Enabling Synthetic Biology Tool," in *Synthetic Biology: Tools and Applications*, ed. Huimin Zhao (Cambridge, MA: Academic Press, 2013), 43–62, https://www.sciencedirect.com/science/article/pii/B9780123944306000030.

32. Clive Cookson, "Sarah Gilbert, the Researcher Leading the Race to a Covid-19 Vaccine," *Financial Times*, July 24, 2020, https://www.ft.com/content/d2ae1fff-9747-41a3-b347-861ac1a70709.

33. Stephanie Baker, "Covid Vaccine Front-Runner Is Months Ahead of Her Competition," Bloomberg, last modified July 15, 2020, https://www.bloomberg.com/news/features/2020-07-15/oxford-s-covid-19-vaccine-is-the-coronavirus-front-runner.

34. "Ebola Vaccine to Begin Human Trials," University of Oxford, November 11, 2021, https://www.ox.ac.uk/news/2021-11-11-ebola-vaccine-begin-human-trials.

35. Carl Zimmer, "Inside Johnson & Johnson's Nonstop Hunt for a Coronavirus Vaccine," *New York Times*, last modified January 13, 2021, https://www.nytimes.com/2020/07/17/health/coronavirus-vaccine-johnson-janssen.html.

36. ClinicalTrials.gov database, www.clinicaltrials.gov (accessed November 3, 2021).

37. Riham Mohamed Aly, "Current State of Stem Cell-Based Therapies: An Overview," *Stem Cell Investigation* 7 (2020): 8, https://pubmed.ncbi.nlm.nih.gov/32695801/.

7. NEW NATURE: A DO-IT-YOURSELF ENVIRONMENT

1. Rebecca Shapiro, author interview via Zoom, March 1, 2021.

2. Robert Carlson, "Estimating the Biotech Sector's Contribution to the US Economy," *Nature Biotechnology* 34, no. 3 (2016): 247–55,

https://www.researchgate.net/publication/297750335_Estimating_the
_biotech_sector%27s_contribution_to_the_US_economy; Rob Carlson
and Rik Wehbring, *Two Worlds, Two Bioeconomies*, Johns Hopkins
Applied Physics Laboratory, 2020, https://www.jhuapl.edu/Content
/documents/Carlson_Wehbring-Biotech.pdf.

3. "The LEGO Group Reveals First Prototype LEGO Brick Made from
Recycled Plastic," LEGO, June 22, 2021, https://www.lego.com/en-us
/aboutus/news/2021/june/prototype-lego-brick-recycled-plastic/.

4. Mark Gillispie, "Billions Pour into Bioplastics as Markets Begin
Ramping Up," AP News, August 9, 2022, https://apnews.com/article
/climate-and-environment-8ddoe4ad3345387ea72e83a506ef3a00.

5. Becky Mackelprang, "Can the Gene Editing Technology Known as
CRISPR Help Reduce Biodiversity Loss Worldwide?" Ensia,
September 13, 2019, https://ensia.com/features/crispr-biodiversity-coral
-food-agriculture-invasive-species/.

6. "GM Crops and Food: GM Crops: Current Situation: Worldwide
Commercial Growing," Gene Watch UK, November 9, 2021, http://
www.genewatch.org/sub-532326.

7. The stories of the Florida panther revival and proposed American chest-
nut gene edit are told in Kent Redford and William Adams, *Strange
Natures: Conservation in the Age of Synthetic Biology* (New Haven, CT:
Yale University Press, 2021).

8. "Frozen Zoo," San Diego Zoo Wildlife Alliance, https://science
.sandiegozoo.org/resources/frozen-zoo%C2%AE (accessed November 3,
2022).

9. "Sentries in the Garden Shed," Homeland Security Website, August 24,
2022, https://www.dhs.gov/sentries-garden-shed.

10. June Medford, presentation at Synthetic Biology: Evolution, Engineering,
and Design Conference (SEED), New York, June 6, 2019.

11. June Medford, et al., "Synthetic Desalination Genetic Circuit in Plants,"
Patent Application, February 2, 13, 2020, https://uspto.report/patent/app
/20200048650.

12. Christina Couch, "The Race Is On to Grow Crops in Seawater and
Feed Millions," Wired UK, June 17, 2020, https://www.wired.co.uk
/article/growing-crops-in-seawater.

13. Warwick Stanley, "This Artificial Mangrove Desalinates Extremely Salty Water," Australian Water Association, February 28, 2020, https://www.awa.asn.au/resources/latest-news/technology/innovation/this-artificial-mangrove-desalinates-water.

14. "How Many People Get Lyme Disease?" Centers for Disease Control, January 13, 2021, https://www.cdc.gov/lyme/stats/humancases.html.

15. A. Burt and R. Trivers, *Genes in Conflict: The Biology of Selfish Genetic Elements* (Cambridge, MA: The Belknap Press of Harvard University Press, 2006).

16. J. Buchthal, S.W. Evans J. Lunshof, S.R. Telford III, K.M. Esvelt, "Mice Against Ticks: An Experimental Community-Guided Effort to Prevent Tick-Borne Disease By Altering the Shared Environment. *Philosophical Transactions of the Royal Society, (*2019) 374: 20180105, http://dx.doi.org/10.1098/rstb.2018.0105

17. Jennifer Kahn, "The Gene-Drive Dilemma: We Alter Entire Species, but Should We?" *New York Times Magazine*, January 8, 2020, https://www.nytimes.com/2020/01/08/magazine/gene-drive-mosquitoes.html.

18. Megan Scudellari, "Self-Destructing Mosquitoes and Sterilized Rodents: The Promise of Gene Drives," *Nature* 571 (2019): 160–62, https://www.nature.com/articles/d41586-019-02087-5.

19. Shapiro, author interview.

20. Claudia Emerson, et. al., "Principles for Gene Drive Research," *Science* 338:6367, December 1, 2017, https://www.science.org/doi/10.1126/science.aap9026.

21. World Health Organization, "Guidance Framework for Testing of Genetically Modified Mosquitoes, Second Edition," 2021, https://www.who.int/publications/i/item/9789240025233.

22. Mamadou Coulibaly, author interview via Zoom, December 16, 2021.

23. "Malaria Fact Sheet," World Health Organization, December 6, 2021, https://www.who.int/news-room/fact-sheets/detail/malaria.

24. "Target Malaria's Gene Drive Project Fails to Inform Local Communities of Risks: New Film," etc Group, December 19, 2018, https://www.etcgroup.org/content/target-malarias-gene-drive-project-fails-inform-local-communities-risks-new-film.

25. Alekos Simoni, author interview via MS Teams, September 14, 2022.

26. Tim Fitzsimmons, "Genetically Engineered Mosquitoes to Be Released in Friday Keys," NBC News, April 17, 2021, https://www.nbcnews.com /science/weird-science/genetically-modified-mosquitoes-be-released -florida-after-years-planning-n1265512.

27. Tim Fitzsimmons, "EPA Oks Plan to Release 2.4 Million More Genetically Modified Mosquitoes," NBC News, March 11, 2022, https://www .nbcnews.com/news/us-news/epa-oks-plan-release-24-million-genetically -modified-mosquitoes-rcna19738.

28. Nathan Rose, author interview via Google Meetings, January 4, 2022.

29. "What Is a Circular Economy?" Environmental Protection Agency, last updated September 29, 2022, https://www.epa.gov/recyclingstrategy /what-circular-economy.

30. Jacob Foss, "Let's Talk About Food Loss," *WholeFoods Magazine*, October 19, 2020, https://wholefoodsmagazine.com/blog/lets-talk-about -food-loss/.

31. John Cumbers, "Synthetic Biology Start-ups Raised $3 Billion in the First Half of 2020," *Forbes*, September 9, 2020, https://www.forbes .com/sites/johncumbers/2020/09/09/synthetic-biology-startups -raised-30-billion-in-the-first-half-of-2020/.

32. George Church and Ed Regis, *Regenesis: How Synthetic Biology Will Reinvent the World and Ourselves* (New York: Basic, 2014).

33. Karsten Temme, quoted in "About Us: Our Story," Pivot Bio, https:// www.pivotbio.com/about-us.

34. "Israeli Wastewater-Electricity Startup Uses Microbial Fuel Cell," Water. Desalination + Reuse, June 29, 2011, https://www.desalination .biz/desalination/israeli-wastewater-electricity-startup-uses-microbial -fuel-cell/.

35. Mia Rozenbaum, "The Increase in Zoonotic Diseases: The WHO, the Why, and the When?" Understanding Animal Research, July 6, 2020, https://www.understandinganimalresearch.org.uk/news/the-increase -in-zoonotic-diseases-the-who-the-why-and-the-when.

36. "Home: Newsroom: Fact Sheets: Zoonoses," World Health Organization, July 29, 2020, https://www.who.int/news-room/fact-sheets/detail /zoonoses.

8. HEARTH AND HOME

1. Chris Carroll, "Growing Beyond Meat," *Maryland Today*, January 17, 2020, https://today.umd.edu/growing-beyond-meat-573f09fb-4847-46ec -bd38-6fa9137242f6.

2. Jonathan McIntyre, presentation at Ginkgo Ferment Conference, Boston, MA, October 28, 2021.

3. "The EVOLUTION of Fake Meat," *Food Network Magazine*, June 2020.

4. Chenelle Bessette, "Ten Questions, Ethan Brown, CEO, Beyond Meat," *Fortune*, January 31, 2014, https://fortune.com/2014/01/31/10-questions -ethan-brown-ceo-beyond-meat/.

5. Tad Friend, "Can a Burger Help Solve Climate Change?" *New Yorker*, September 30, 2019, 44, https://www.newyorker.com/magazine/2019 /09/30/can-a-burger-help-solve-climate-change.

6. Florian Humpenöder, Benjamin. L. Bodirsky, Iisabelle Weindl, et al., "Projected Environmental Benefits of Replacing Beef with Microbial Protein," *Nature* 605 (2022): 90–96, https://doi.org/10.1038/s41586-022 -04629-w.

7. Bridgit Bowden, "How We Produce More Milk with Fewer Cows: Breeding and DNA Contribute to Rising Milk per Cow," Wisconsin Public Radio, March 28, 2017, https://www.wpr.org/how-we-produce -more-milk-fewer-cows.

8. Rachael Link, "A Dietitian Reviews Taste and Nutrition of the Impossible Burger," *Healthline*, September 22, 2022, https://www.healthline .com/nutrition/impossible-burger; "Listen to that Burger Sizzle," Twitter, https://twitter.com/impossiblefoods/status/1555255421438402560 ?lang=ca

9. Laura Reiley, "Impossible Burger: Here's What's Really in It," *Washington Post*, October 23, 2019, https://www.washingtonpost.com/business /2019/10/23/an-impossible-burger-dissected/.

10. "What Is Lab Grown Meat?" GCF Global, https://edu.gcfglobal.org /en/thenow/what-is-labgrown-meat/1/ (accessed November 4, 2022).

11. Michael Jewett, quoted in Monique Brouillette, "Repurposing the Cell Engine: Northwestern Announces the Cell-Free Biomanufacturing Institute," *GEN Biotechnology* 1, no. 3 (June 14, 2022), https://www .liebertpub.com/doi/full/10.1089/genbio.2022.29037.mbr.

12. Claire Toenisketter, "Lab-Grown Meat Receives Clearance from FDA," *New York Times*, November 17, 2022, https://www.nytimes.com/2022/11/17/climate/fda-lab-grown-cultivated-meat.amp.html.

13. Andy Coyne, "Eyeing Alternatives—Meat Companies with Stakes in Meat-Free And Cell-Based Meat," Just Food, November 21, 2021, https://www.just-food.com/features/eyeing-alternatives-meat-companies-with-stakes-in-meat-free-and-cell-based-meat/.

14. Oliver Garret, "Why Bill Gates Is Betting Millions on Synthetic Biology," *Forbes*, September 10, 2020, https://www.forbes.com/sites/oliviergarret/2020/09/10/why-bill-gates-is-betting-millions-on-synthetic-biology/?sh=7bc4857765c6.

15. Cecil W. Forsberg, Roy G. Meidinger, Mingfu Liu, Michael Cottrill, Serguei Golovan, John P. Phillips, "Integration, Stability, and Expression of the *E. coli* Phytase Transgene in the Cassie Line of Yorkshire Enviropig," *Transgenic Research*, 2013 April; 22, no. 2: 379–89.

16. Stephanie Michelsen, presentation at SynBioBeta Agriculture Zoom Conference, April 2020, www.synbiobeta.com.

17. Elizabeth Green, "Jellatech Exec Eyes Growth for Animal-Free Gelatin as Scale-up Is on the Horizon," Food Ingredients First, May 11, 2021, https://www.foodingredientsfirst.com/news/jellatech-exec-eyes-growth-for-animal-free-collagen-and-gelatin-as-scale-up-is-on-the-horizon.html.

18. Dan Widmaier, author interview via Zoom. December 13, 2021

19. Stella McCartney, quoted in Elizabeth Paton, "Fungus May Be Fall's Hottest Fashion Trend," *New York Times*, last modified October 5, 2020, https://www.nytimes.com/2020/10/02/fashion/mylo-mushroom-leather-adidas-stella-mccartney.html.

20. Eben Bayer, quoted in Becca Tucker, "The Making of a Biotech Pioneer," Dirt, July 2, 2018, https://www.dirt-mag.com/stories/the-making-of-a-biotech-pioneer-ARDM20180702180709995.

21. Eben Bayer, "Grow," by Ecovative Design, YouTube, September 19, 2013 [video], 1:21–1:25, https://youtube/Zh-2TEmLtQE.

22. Michael Seldon, presentation at SynBioBeta Food & Agriculture Zoom Conference, March 5, 2021, www.synbiobeta.com.

23. Biodesign Challenge 2022, https://www.biodesignchallenge.org/summit-2022 (accessed November 4, 2022).

24. Mustafa Germec, Ali Demirci, and Irfan Turhan, "Biofilm Reactors for Value-Added Products Production: An In-depth Review," *Biocatalysis and Agricultural Biotechnology* 27 (August 2020): 101662, https://www.sciencedirect.com/science/article/abs/pii/S1878818119313568.

25. Tom Adams, author interview via Zoom, March 12, 2021.

26. Mikhaela Neequaye, "Genetic Control of Sulphur Metabolism in Broccoli: How to Make Better Broccoli," UK Research and Innovation, https://gtr.ukri.org/projects?ref=studentship-1653460#/tabOverview (accessed November 4, 2022).

27. "Synthetic Biology Global Market Report 2022," GlobeNewsWire, February 18, 2022, https://www.globenewswire.com/news-release/2022/02/18/2388033/0/en/Synthetic-Biology-Global-Market-Report-2022.html.

28. Benjamin Ferrer, "Givaudan & Ginkgo Bioworks Collaboration Eyes Rare, Complex and Natural Ingredients for NPD," Food Ingredients First, August 31, 2021, https://www.foodingredientsfirst.com/news/givaudan-ginkgo-bioworks-collaboration-eyes-rare-complex-and-natural-ingredients-for-sustainable-npd.html.

9. FANTASTIC VOYAGES: MINING AND THE MILITARY

1. J. Gentina and F. Acevedo, "Copper Bioleaching in Chile," *Minerals* 6, no. 1 (March 2016): 23, https://doi.org/10.3390/min6010023.

2. Lindsey Delevingne, Will Glazener, Liesbel Gregoir, Kimberly Henderson, "Vlimare Risk and Decarbonization: What Every Mining CEO Needs to Know," *McKinsey Sustainability*, January 28, 2020, https://www.mckinsey.com/capabilities/sustainability/our-insights/climate-risk-and-decarbonization-what-every-mining-ceo-needs-to-know; T. Kuykendall, K. Bouckley, S. Tsao, and G. Doulakia, "Mining Faces Pressure for Net-Zero Carbon Emissions, As Demand Rises for Clean Energy," *S&P Global Commodity Insights, July 17, 2020*, https://www.spglobal.com/commodityinsights/en/market-insights/latest-news/coal/072720-mining-faces-pressure-for-net-zero-targets-as-demand-rises-for-clean-energy-raw-materials.

3. Katia Moskvitch, "Biomining: How Microbes Help to Mine Copper," BBC News, March 21, 2012, https://www.bbc.com/news/technology -17406375.

4. Kai Kupperschmidt, "A Paper Showing How to Make a Smallpox Cousin Just Got Published. Critics Wonder Why," *Science*, January 19, 2018, https://www.science.org/content/article/paper-showing-how-make -smallpox-cousin-just-got-published-critics-wonder-why.

5. Q. Wang, B. Kille, T. R. Liu, et al., "PlasmidHawk Improves Lab of Origin Predictions of Engineered Plasmids Using Sequence Alignment," *Nature Communications* 12 (2021), https://www.nature.com/articles /s41467-021-21180-w.pdf.

6. Panel Discussion, "Care at Ginkgo: An Expert Conversation About Biosecurity," Ginkgo Investor's Day, July 8, 2021, https://www.youtube.com /watch?v=PIX3lNSz9JM.

7. George Church, author interview by phone, May 15, 2019.

8. Friedrich Frischknecht, "The History of Biological Warfare," in *Decontamination of Warfare Agents*, ed. André Richardt and Marc-Michael Blum (Weinheim: Wiley-VCH, 2008), https://application.wiley-vch .de/books/sample/3527317562_c01.pdf.

9. Patrick Kiger, "Did Colonists Give Infected Blankets to Native Americans as Biological Warfare?" History.com, updated November 25, 2019, https:// www.history.com/news/colonists-native-americans-smallpox-blankets.

10. Michelle Rozo, Gigi Kwik Gronvall, "The Reemergent 1977 H1N1 Strain and the Gain-of-Function Debate," *mBio*, July/August, 2015, 6(4): e01013-15, https://www.ncbi.nlm.nih.gov/pmc/articles/PMC4542197/.

11. Martin Furmanski, "Laboratory Escapes and 'Self-Fulfilling Prophecy' Epidemics," Center for Arms Control and Non-Proliferation, February 17, 2014, https://armscontrolcenter.org/wp-content/uploads/2016/02 /Escaped-Viruses-final-2-17-14-copy.pdf.

12. Story told in Matt Ridley and Alina Chan, *Viral: The Search for the Origin of COVID-19* (New York: Harper, 2021), 143; J. Wertheim, "The Re-Emergence of H1N1 Influenza Virus in 1977: A Cautionary Tale for Estimating Divergence Times Using Biologically Unrealistic Sampling Dates," *Public Library of Science* 5, no. 6 (June 2010): e11184, https:// doi.org/10.1371/journal.pone.0011184.

13. "The 1979 Anthrax Leak in Sverdlovsk," *Frontline*, PBS, https://www.pbs .org/wgbh/pages/frontline/shows/plague/sverdlovsk/.

14. Ken Alibek, *BioHazard: The Chilling True Story of the Largest Undercover Biological Weapons Program in the World—Told from the Inside by the Man Who Ran It* (New York: Random House, 1999).

15. A. C. Brault, A. M. Powers, G. Medina, et al., "Potential Source of the 1995 Venezuelan Equine Encephalitis Subtype 1C Epidemic," *Journal of Virology* 75, no. 13 (July 1, 2001), https://journals.asm.org/doi/full/10.1128/JVI.75.13.5823-5832.2001.

16. Jane Parry, "Breaches of Safety Regulations Are Probable Cause of Recent SARS Outbreak, WHO Says." *BMJ* 328, no. 7450 (May 22, 2004), https://www.ncbi.nlm.nih.gov/pmc/articles/PMC416634/.

17. Iain Andersen, *Foot and Mouth Disease 2007: A Review*, Report Presented to the Prime Minister and the Secretary of State for Environment, Food and Rural Affairs, London: 2007, https://assets.publishing.service.gov.uk/government/uploads/system/uploads/attachment_data/file/250363/0312.pdf.

18. Jessie Yeung and Eric Cheung, "Bacterial Outbreak Infects Thousands After Factory Leak in China," CNN, September 17, 2020, https://www.cnn.com//2020/09/17/asia/china-brucelllosis-outbreak-intl-hnl/index.html.

19. Juan Carlos Gentina, Fernando Acevedo, "Copper Bioleaching in Chile," *Minerals*, 6:23, March 16, 2016, https://www.mdpi.com/2075-163X/6/1/23

20. See, for instance, "Researchers Gain Insights into Rio Tinto Microbial Community," Phys.org, February 23, 2022, https://phys.org/news/2022-02-gain-insights-rio-tinto-microbial.html.

21. "What Is Biomining?" American Geosciences Institute, https://www.americangeosciences.org/critical-issues/faq/what-biomining (accessed November 26, 2021).

22. Pilar Parada Valdecantos, quoted in Katia Moskvitch, "Biomining: How Microbes Help To Mine Copper," BBC News, March 21, 2021, https://www.bbc.com/news/technology-17406375

23. Indrani Banerjee, Brittany Burrell, Cara Reed, et al., "Metals and Minerals as a Biotechnology Feedstock: Engineering Biomining Microbiology for Bioenergy Applications," *Current Opinion in Biotechnology* 45 (June 2017): 144–55, https://www.sciencedirect.com/science/article/abs/pii/S0958166916302403.

24. Farhan Mita, "Biomining: Turning Waste into Gold With Microbes," LabBiotech.eu, January 18, 2022, https://www.labiotech.eu/in-depth/biomining-sustainable-microbes/.

25. Plantguy, "Plant Detectorists, Part 2—Using Plants to Prospect for Gold," How Plants Work, November 5, 2020, https://www.howplants work.com/2020/11/05/plant-detectorists-part-2-using-plants-to-prospect -for-gold/.

26. "Bioleaching: Making Mining Sustainable," *MIT Mission 2015: Biodiversity*, https://web.mit.edu/12.000/www/m2015/2015/bioleaching.html; M. Capeness and L. Horsfall, "Synthetic Biology Approaches Toward the Recycling of Metals from the Environment," *Biochemical Society Transactions* 48, no. 4 (July 2020): 1367–78, https://doi.org/10.1042/BST20190837.

27. Heidi Shyu, quoted in "New Biotechnology Executive Order Will Advance DoD's Biotechnology Initiatives for America's Economic and National Security," September 14, 2022, https://www.defense.gov /News/Releases/Release/Article/3157504/new-biotechnology-executive -order-will-advance-dod-biotechnology-initiatives-fo/.

28. Nancy Kelley-Loughnane, "Future of Material Synthesis for Military Environments," SEED Conference, June 25, 2019.

29. "Synthetic Biology," Air Force Research Laboratory, https://afresearchlab .com/technology/human-performance/synthetic-biology.

30. Yvonne Linney, quoted in "Automation for the People," Babbage from *The Economist*, February 28, 2018, https://pca.st/n6EI.

31. "Bioremediation of Dumped Waste," PIB India, YouTube, May 17, 2018 [video], https://www.youtube.com/watch?v=4pTqFd5upFA.

32. Scott Banta, presentation at SEED conference, New York, NY, June 6, 2019.

33. Lt. Col Marcus Cunningham, USAF, "A National Strategy for Synthetic Biology," *Strategic Studies Quarterly* 14, no. 3 (Fall 2020): 49–80, https:// www.jstor.org/stable/pdf/26937411.pdf?refreqid=excelsior%3Ab22bfe0a 09a8dea56767223c7124724d&ab_segments=&origin=.

34. Sarah Gilbert, "COVID Conversations," Oxford University, June 17, 2020, https://www.youtube.com/watch?v=MKNavonhXyk.

35. Jennifer Doudna, quoted in Michael Chui, "Programming Life: An Interview with Jennifer Doudna," McKinsey & Company, June 4, 2020, https://www.mckinsey.com/industries/life-sciences/our-insights /programming-life-an-interview-with-jennifer-doudna.

36. Edward Regis and George Church, *Regenesis: How Synthetic Biology Will Reinvent Nature and Ourselves* (New York: Basic, 2014), 235.

37. Church, author interview.

10. THE KILLERS: VIRUSES AS HEALERS

1. Julianne Lemieux, "Q-and-A, Graham Hatfull," *Phage: Therapy, Application and Research* 1, no. 1 (2020): 4–9.

2. "How Phage Therapy Kills Superbugs: Weaponizing Viruses to Fight Infections," Yale Medicine, November 18, 2019, https://www.yale medicine.org/stories/phage-therapy/.

3. Dimitry Melnikov, "Creature Features: The Lively Narratives of Bacteriophages in Soviet Biology and Medicine," *The Royal Society Journal of the History of Science*, January 15, 2020, https://royalsocietypublishing.org/doi/10.1098/rsnr.2019.0035.

4. Author interview via Zoom, Graham Hatfull, September 22, 2021.

5. James Tobin, "The Michigan Scientist Who Was Arrowsmith," The Heritage Project, University of Michigan, ttps://heritage.umich.edu/stories/the-michigan-scientist-who-was-arrowsmith/.

6. Steffanie Strathdee, author interview by phone, May 24, 2021.

7. Hatfull, author interview.

8. Charles Schmidt, "Phage Therapy's Latest Makeover," *Nature Biotechnology* 37 (2019): 583.

9. Hatfull, author interview.

10. Daniela I. Staquicini, Fenny H. F. Tang, Christopher Markosian, et al., "Design and Proof of Concept for Targeted Phage-Based COVID-19 Vaccination Strategies with a Stream-Lined Cold-Free Supply Chain," *Proceedings of the National Academy of Sciences USA* 118, no. 30 (July 7, 2021): e2105739118, https://www.pnas.org/content/118/30/e2105739118.

11. Anca Segall, author interview via Zoom, May 18, 2021.

12. Rob Stein, "Genetically Modified Viruses Help Save A Patient with a 'Superbug' Infection," *All Things Considered*, NPR, May 8, 2019, https://www.npr.org/sections/health-shots/2019/05/08/719650709/genetically-modified-viruses-help-save-a-patient-with-a-superbug-infection.

13. Ahlam Alsaadi, Beatriz Beamud, Maheswaran Easwaran, et al., "Learning from Mistakes: The Role of Phages in Pandemics," *Frontiers in Microbiology* 12 (March 17, 2021): 653107, https://www.frontiersin.org/articles/10.3389/fmicb.2021.653107/full.

14. Louise Temple and Lynn Lewis, "Phage on the Stage" *Bacteriophage* 5, no. 3 (June 22, 2015): e1062589.

15. Paul Turner, author interview via Zoom, June 9, 2021.

16. Schmidt, "Phage Therapy's Latest Makeover," 586.

17. Strathdee, author interview.

18. Graham Hatfull, e-mail to author, October 16, 2021.

19. Chris Dahl, "Scientists, Biotechs Look to Unlock the Potential of Phage Therapy," CIDRAP News, October 5, 2021, https://www.cidrap.umn.edu /news-perspective/2021/10/scientists-biotechs-look-unlock-potential -phage-therapy.

20. Paul Turner, author interview via Zoom, June 9, 2021

21. Antonio Regalado and Neel Patel, "Mammoth Biosciences," *MIT Technology Review*, May 20, 2020.

22. Strathdee, author interview.

23. Mallory Smith, *Salt in My Soul: An Unfinished Life* (New York: Random House), 2019.

24. Strathdee, author interview.

25. Turner, author interview.

26. Ella Balasa, author interview via Zoom, June 11, 2021.

27. PhagePro, "PhagePro Receives $3.1 Million in Funding from the National Institutes of Health," Press Release, March 11, 2021, https:// phageproinc.com/press-releases.

28. Graham Hatfull, email to author, January 6, 2023

29. Steffanie A. Strathdee, Graham F. Hatfull, Vivek K. Mutalik, Robert T. Schooley, "Phage Therapy: From Biological Mechanisms to Future Directions." *Cell*, 186 (January 5, 2023): 17–31, https://www.cell.com/cell /pdf/S0092-8674(22)01461-1.pdf.

11. RACE TO A VACCINE

1. Damian Garde, "Lavishly Funded Moderna Hits Safety Problems in Bold Bid to Revolutionize Medicine," STAT, January 10, 2017, https:// www.statnews.com/2017/01/10/moderna-trouble-mrna/.

2. Kelly Servick, "This Mysterious $2 Billion Biotech is Revealing the Secrets Behind Its New Drugs and Vaccines," *Science*, March 25, 2020, https://www.science.org/content/article/mysterious-2-billion-biotech -revealing-secrets-behind-its-new-drugs-and-vaccines.

3. Denise Chow, "What Is mRNA? How Pfizer and Moderna Tapped New Tech to Make Coronavirus Vaccines," NBC News, November 17,

2020, https://www.nbcnews.com/science/science-news/what-mrna-how
-pfizer-moderna-tapped-new-tech-make-coronavirus-n1248054.

4. Charlie Campbell, "Exclusive: The Chinese Scientist Who Sequenced
   the First COVID-19 Genome Speak Outs About the Controversies
   Surrounding His Work," *Time*, August 24, 2020, https://time.com
   /5882918/zhang-yongzhen-interview-china-coronavirus-genome/.

5. M. Cascella, M. Rajnik, A. Aleem, et al., "Features, Evaluation, and
   Treatment of Coronavirus," StatPearls, last updated October 13, 2022,
   https://www.ncbi.nlm.nih.gov/books/NBK554776/.

6. Overview of COVID-19 Vaccines," CDC, updated November 1, 2022,
   https://www.cdc.gov/coronavirus/2019-ncov/vaccines/different-vaccines
   /Pfizer-BioNTech.html.

7. Stephanie Soucheray, "Coroner: First US COVID-19 Death Occurred
   in Early February," Center for Infectious Disease Research and
   Policy (CIDRAP), April 22, 2020, https://www.cidrap.umn.edu/news
   -perspective/2020/04/coroner-first-us-covid-19-death-occurred-early
   -february.

8. Benjamin Neuman, quoted in Brooke Jarvis, "The First Shot: Inside
   the Covid Vaccine Fast Track," *Wired*, May 13, 2020, https://www.wired
   .com/story/moderna-covid-19-vaccine-trials/.

9. Ryan Cross, "Adenoviral Vectors Are the New COVID-19 Vaccine
   Front-Runners. Can They Overcome Their Checkered Past?" *Chemical and
   Engineering News*, May 12, 2020, https://cen.acs.org/pharmaceuticals
   /vaccines/Adenoviral-vectors-new-COVID-19/98/i19.

10. John Cumbers, "The Synthetic Biology Companies Racing to Fight
    Coronavirus," *Forbes*, February 5, 2020, https://www.forbes.com/sites
    /johncumbers/2020/02/05/seven-synthetic-biology-companies-in-the
    -fight-against-coronavirus/?sh=6d7c51a316ef.

11. David Gelles, "The Husband and Wife Team Behind the Leading Vac-
    cine to Solve COVID-19," *New York Times*, Nov. 10, 2020, D1, https://
    www.nytimes.com/2020/11/10/business/biontech-covid-vaccine
    .html

12. Scott Gottlieb, presentation at the Ginkgo Ferment 2021 Conference,
    Boston, MA, October 28, 2021.

13. Sarah Gilbert, "Racing Against the Virus," Rosalind Franklin Lecture,
    2021. March 5, 2021, https://www.youtube.com/watch?v=U29oFk2BpvA.

14. "Pfizer and BioNTech Dose First Participants in the US as Part of a Global COVID-19 MRNA-Based Vaccine Program" [press release], Pfizer News, May 5, 2020, https://www.pfizer.com/news/press-release/press -release-detail/pfizer_and_biontech_dose_first_participants_in_the_u_s _as_part_of_global_covid_19_mrna_vaccine_development_program.

15. "Moderna and Lonza Announce Worldwide Strategic Collaboration to Manufacture Moderna's Vaccine (mRNA-1273) Against Novel Coronavirus" [press release], Lonza News, May 1, 2020, https://www .lonza.com/news/2020-05-01-04-50.

16. "A Clinical Trial of a COVID-19 Vaccine in Healthy Adults," Moderna Clinical Trials, https://trials.modernatx.com/study/?id=mRNA-1273-P301 (accessed November 7, 2022).

17. Caroline Copley and Thomas Seythal, "Germany Grants BioNTech, CureVac $745 Million to Speed Up COVID-19 Vaccine Work," Reuters, September 15, 2020, https://www.reuters.com/article/us-health-coronavirus -germany-vaccine/germany-grants-biontech-curevac-745-million-to -speed-up-covid-19-vaccine-work-idUSKBN2661JB.

18. Elie Dolgin, "CureVac COVID Vaccine Let-Down Spotlights mRNA Design Challenges," Nature, June 18, 2021, https://www.nature.com /articles/d41586-021-01661-0.

19. Kyle Blankenship, "Johnson and Johnson Sets Stage for Vaccine Rollout with 'First in-a-Series' Manufacturing Deal," FiercePharma, April 24, 2020, https://www.fiercepharma.com/manufacturing/johnson-johnson -sets-stage-for-covid-19-vaccine-rollout-emergent-manufacturing-tie-up.

20. Jonathan Corum and Carl Zimmer, "How the Johnson & Johnson Vac- cine Works," New York Times, May 7, 2021, https://www.nytimes.com /interactive/2020/health/johnson-johnson-covid-19-vaccine.html.

21. Julie Steenhuysen, "J&J, Moderna Sign Deals with U.S. to Produce Huge Quantity of Possible Coronavirus Vaccines," Reuters, March 30, 2022, https://www.reuters.com/article/us-health-coronavirus-johnson -johnson/jj-moderna-sign-deals-with-u-s-to-produce-huge-quantity -of-possible-coronavirus-vaccines-idUSKBN21H1OY.

22. CanSino Biologics, "A Best Shot at Global Public Health Response" [advertisement feature], Nature, https://www.nature.com/articles/d42473 -018-00219-5 (accessed November 7, 2022).

23. "COVID Shot Drives CanSinoBIO's First Six-Month Profit Since at Least 2019," Reuters, August 27, 2021, https://money.usnews.com

/investing/news/articles/2021-08-27/covid-shot-drives-cansinobios
-first-six-month-profit-since-at-least-2019.

24. F. C. Zhu, Y. H. Li, X. H. Guan, et al., "Safety, Tolerability, and Immu-
nogenicity of a Recombinant Adenovirus Type-5 Vectored COVID-19
Vaccine: A Dose-Escalation, Open-Label, Non-Randomized, First-in
-Human Trial," *The Lancet* 395, no. 10240 (June 2020): 1845–54,
https://doi.org/10.1016/S0140-6736(20)31208-3.

25. L. F. Wang and B. T. Eaton, "Bats, Civets, and the Emergence of
SARS," *Current Topics in Microbiological Immunology* 315 (2007): 325–44,
https://www.ncbi.nlm.nih.gov/pmc/articles/PMC7120088/pdf/978-3
-540-70962-6_Chapter_13.pdf.

26. Ben Hu, Lei-Ping Zeng, Xing-Lou Yang, et al., "Discovery of a Rich
Gene Pool of Bat SARS-Related Coronaviruses Provides New Insights
into the Origin of SARS Coronaviruses," *PLoS Pathogens*, November 30,
2017, https://journals.plos.org/plospathogens/article?id=10.1371/journal
.ppat.1006698; David Cyranoski, "Bat Cave Solves Mystery of Deadly
SARS Virus—and Suggests New Outbreak Could Occur," *Nature* 552
(December 1, 2017): 15–16, https://www.nature.com/articles/d41586-017
-07766-9.

27. Alina Chan and Matt Ridley, *Viral: The Search for the Origin of COVID-
19* (New York: Harper, 2021), 203.

28. Nikolai Petrovsky, quoted in Charles Schmidt, "Lab Leak: A Scientific
Debate Mired in Politics—and Unresolved," *Undark*, March 17, 2021,
https://undark.org/2021/03/17/lab-leak-science-lost-in-politics/.

29. "Mission Summary: WHO Field Visit to Wuhan, China 20-21 January
2020," World Health Organization, January 22, 2020, https://www.who
.int/china/news/detail/22-01-2020-field-visit-wuhan-china-jan-2020.

30. Nicholas Wade, "The Origin of COVID: Did People or Nature Open
Pandora's Box at Wuhan?" *Bulletin of Atomic Scientists*, May 5, 2021,
https://thebulletin.org/2021/05/the-origin-of-covid-did-people-or
-nature-open-pandoras-box-at-wuhan/.

31. Chan and Ridley, *Viral*, 69.

32. Tao Zhang, Qunfu Wu, and Zhigang Zhang, "Probable Pangolin Origin of
SARS-CoV-2 Associated with the COVID-19 Outbreak," *Current Biology*
30, no. 7 (2020): 1346–51, https://pubmed.ncbi.nlm.nih.gov/32197085/.

33. Ping Liu, Wu Chen, and Jin-Ping Chen, "Viral Metagenomics Revealed
Sendai Virus and Coronavirus Infection of Malayan Pangolins (*Manis*

javanica)," *Viruses* 11, no. 11 (October 24, 2019): 979, https://www.mdpi .com/1999-4915/11/11/979.

34. Michelle Fay Cortez, "The Last-And Only-Foreign Scientist in the Wuhan Lab Speaks Out." Bloomberg News, (June 27, 2021) https:// www.bloomberg.com/news/features/2021-06-27/did-covid-come -from-a-lab-scientist-at-wuhan-institute-speaks-out#xj4y7vzkg

35. David Stanway, "Explainer: China's Mojiang Mine and Its Role in the Origins of COVID-19," Reuters, June 9, 2021, https://www.reuters.com /business/healthcare-pharmaceuticals/chinas-mojiang-mine-its-role -origins-covid-19-2021-06-09/.

36. Chan and Ridley, *Viral*, 9–31.

37. Director of National Intelligence "Unclassified Summary of Find- ings on COVID-19 Origin," October 29, 2021, https://www.dni.gov /files/ODNI/documents/assessments/Declassified-Assessment-on -COVID-19-Origins.pdf.

38. Lori Robertson, "The Wuhan Lab and the Gain-of-Function Dis- agreement," FactCheck.org, May 21, 2021, https://www.factcheck.org /2021/05/the-wuhan-lab-and-the-gain-of-function-disagreement/.

39. Antonio Regalado, @antonioregalado Tweet, September 7, 2021, https:// twitter.com/antonioregalado/status/1435317759718604808?s=11.

40. Rowan Jacobsen, "Inside the Risky Bat-Virus Engineering That Links America to Wuhan, *MIT Technology Review*, June 29, 2021, https:// www.technologyreview.com/2021/06/29/1027290/gain-of-function -risky-bat-virus-engineering-links-america-to-wuhan/.

41. Helen Branswell, "NIH Awards $7.5 Million Grant to EcoHealth Alliance, Months After Uproar Over Political Interference," STAT, August 27, 2020, https://www.statnews.com/2020/08/27/nih-awards -grant-to-ecohealth-alliance-months-after-uproar-over-political -interference/.

42. Carl Zimmer, "Newly Discovered Bat Viruses Give Hints to Covid's Origins," *New York Times*, October 19, 2021, https://www.nytimes.com /2021/10/14/science/bat-coronaviruses-lab-leak.html.

43. Robert M. Beyer, Andrea Manica, and Camilo Mora, "Shifts in Global Bat Diversity Suggest a Possible Role of Climate Change in the Emer- gence of SARS-CoV-1 and SARS-CoV-2," *Science of the Total Environ- ment* 767 (May 1, 2021): 145413, https://www.sciencedirect.com/science /article/pii/S0048969721004812?via%3Dihub.

44. Michael Worobey, Joshua I. Levy, Lorena Malpica Serrano, et al., "The Huanan Seafood Wholesale Market in Wuhan Was the Early Epicenter of the COVID-19 Pandemic," *Science* 377, no. 6609 (July 26, 2022): 951–59, https://www.science.org/doi/10.1126/science.abp8715.

45. Jon Cohen, "Wuhan Coronavirus Hunter Zhi Zheng-Li Speaks Out," *Science* 369 (July 30, 2020): 6503, 487–88, https://www.science.org/doi/10.1126/science.369.6503.487.

46. There are no new interviews after 2021, and she told the *New York Times* she was "calming down": Amy Qin, Chris Buckley, "A Top Virologist in China, At Center Of a Pandemic Storm, Speaks Out," *New York Times*, June 14, 2021, https://www.nytimes.com/2021/06/14/world/asia/china-covid-wuhan-lab-leak.html.

47. "Spotlight: Chinese Scientist Tells *Science* Origin of COVID-19 Undetermined," July 28, 2020, http://www.xinhuanet.com/english/2020-07/28/c_139246494.htm.

48. Shi Zheng-Li, "Shi Zheng-Li Post SARS-CoV-2 in *Science* Magazine," Covid-19 Research Institute, March 2020, https://covid19researchinstitute.com/#post.

49. Will Canine, author interview via Zoom, November 4, 2021.

50. "Charles River Labs Acquires Antibody Partner Distributed Bio for Up to $104M," *Genetic Engineering and Biotechnology News*, January 5, 2021, https://www.genengnews.com/news/charles-river-labs-acquires-antibody-partner-distributed-bio-for-up-to-104m/.

51. Argonne National Laboratory, "COVID-19 Research: Argonne Scientists Help Track, Treat and Stop the Spread of the Global Pandemic," Argonne National Laboratory, https://www.anl.gov/coronavirus (accessed December 5, 2022).

52. J. Collins, et al., "Wearable Materials With Embedded Synthetic Biology Sensors For BioMolecule Detection," *Nature Biotechnology* 39: 1366-1374 (2021) June 28, 2021, https://www.nature.com/articles/s41587-021-00950-3.

12. GLOBAL PRODUCTION: PERILS AND PROFITS OF A NEW SCIENCE

1. Feng Li, Natarajan Vijayasankaran, Amy (Yijuan) Shen, et al., "Cell Culture Processes for Monoclonal Antibody Production," *MAbs* 2, no. 5 (2010): 466–79, https://www.ncbi.nlm.nih.gov/pmc/articles/PMC2958569/.

2. "Global COVID-19 Death Toll Tops 2 Million," Reuters, January 15, 2021, https://www.reuters.com/article/us-health-coronavirus-global -casualties/global-covid-19-death-toll-tops-2-million-idUSKBN29K2EW.

3. "David Adam, "COVID's True Death Toll: Much Higher than Official Records," *Nature*, March 10, 2022, https://www.nature.com/articles /d41586-022-00708-0.

4. "Black Death," History.com, https://www.history.com/topics/middle -ages/black-death#:~:text=Sicilian%20authorities%20hastily %20ordered%20the,third%20of%20the%20continent's%20population (accessed November 8, 2022).

5. George Yancopoulos, quoted in "Can A Vaccine for Covid-19 Be Developed in Record Time? A Discussion Moderated by Siddhartha Mukherjee," *New York Times Magazine*, June 14, 2020, 38.

6. "Pfizer and BioNTech Celebrate First Authorization in the US of Vaccine to Prevent COVID-19," Press release, December 11, 2020, https://www.pfizer.com/news/press-release/press-release-detail /pfizer-and-biontech-celebrate-historic-first-authorization.

7. Emma Cott, Elliot deBruyn, and Jonathan Corum, "How Pfizer Makes Its Covid-19 Vaccine," *New York Times*, April 28, 2021, https://www .nytimes.com/interactive/2021/health/pfizer-coronavirus-vaccine.html.

8. "Pfizer Facility in Chesterfield Scaling Up Operations for Production of COVID-19 Vaccine," Governor's Press Release, May 7, 2020, https:// governor.mo.gov/press-releases/archive/pfizer-facility-chesterfield -missouri-scaling-operations-covid-19-vaccine; https://www.nytimes .com/interactive/2021/health/pfizer-coronavirus-vaccine.html.

9. Christopher Rowland, "Inside Pfizer's Race to Produce the World's Biggest Supply of Coronavirus Vaccine," *Washington Post*, June 16, 2021, https://www.washingtonpost.com/business/2021/06/16/pfizer-vaccine -engineers-supply/.

10. Ryan Cross, "Without These Lipid Shells, There Would Be No mRNA Vaccines for COVID-19," *Chemical and Engineering News*, March 6, 2021, https://cen.acs.org/pharmaceuticals/drug-delivery/Without-lipid -shells-mRNA-vaccines/99/i8.

11. "Shot of a Lifetime: How Two Pfizer Manufacturing Plants Upscaled to Produce the COVID-19 Vaccine in Record Time," Pfizer, https:// www.pfizer.com/news/articles/shot_of_a_lifetime_how_two_pfizer _manufacturing_plants_upscaled_to_produce_the_covid_19_vaccine _in_record_time (accessed November 8, 2022).

12. Zain Rizvi, Jishian Ravinthiran, and Amy Kapczynski, "Sharing The Knowledge: How President Joe Biden Can Use The Defense Production Act To End The Pandemic Worldwide," Health Affairs, August 6, 2021, https://www.healthaffairs.org/do/10.1377/forefront.20210804 .101816/.

13. Christopher Rowland, "Inside Pfizer's Race to Produce the World's Biggest Supply of Covid Vaccine," *Washington Post*, June 16, 2021, https:// www.washingtonpost.com/business/2021/06/16/pfizer-vaccine -engineers-supply/.

14. "COVID-19," Ginkgo Bioworks, https://www.ginkgobioworks.com /covid-19/ (accessed December 8, 2021).

15. "Products: SARS-CoV-2 Tools: Overview: Tools for Addressing the SARS-CoV-2 Virus," Twist Bioscience, https://www.twistbioscience.com /productchild/tools-addressing-sars-cov-2-virus (accessed December 8, 2021).

16. Antonio Regalado, "Biologists Rush to Re-Create the China Coronavirus from Its Genetic Code," *MIT Technology Review*, February 15, 2020, https://www.technologyreview.com/2020/02/15/844752/biologists -rush-to-re-create-the-china-coronavirus-from-its-dna-code/.

17. Rebecca Mackelprang, Katarzyna P. Adamala, Emily R. Aurand, et al., "Making Security Viral: Shifting Engineering Biology Culture and Publishing," *ACS Synthetic Biology* 11, no. 2 (2022): 522–27, https://pubs .acs.org/doi/10.1021/acssynbio.1c00324#.Yg-SGCLo4zp.linkedin.

18. "Sherlock Biosciences' 221b Foundation Enables Global Access to COVID-19 Diagnostics with Up to 10 Million Tests per Month," Sherlock Biosciences, June 22, 2021, https://sherlock.bio/sherlock -biosciences-221b-foundation-enables-global-access-to-covid-19 -diagnostics-with-up-to-10-million-tests-per-month/.

19. "Global Tissue Engineering Market to Reach $26.8 Billion by 2027," GlobeNewswire, April 21, 2021, https://www.globenewswire.com/news -release/2021/04/21/2214287/0/en/Global-Tissue-Engineering-Market -to-Reach-26-8-Billion-by-2027.html.

20. Carlos del Rio, "Why Monoclonal Antibody COVID Therapies Have Not Lived Up to Expectations," *Scientific American*, May 18, 2021, https:// www.scientificamerican.com/article/why-monoclonal-antibody-covid -therapies-have-not-lived-up-to-expectations/.

21. John C. O'Horo, et al., "Effectiveness of Monoclonal Antibodies in Preventing Severe COVID-19 with Emergence of the Delta Variant,"

Mayo Clinic Proceedings, February 2022, 97 (2): 327–32, https://www
.mayoclinicproceedings.org/article/S0025-6196(21)00924-1/fulltext.

22. "Coronavirus (COVID-19) Update: FDA Revokes Emergency Use
Authorization for Monoclonal Antibody Bamlanivimab," FDA, April 16,
2021, https://www.fda.gov/news-events/press-announcements/coronavirus
-covid-19-update-fda-revokes-emergency-use-authorization-monoclonal
-antibody-bamlanivimab.

23. Rebecca Robbins, "Pfizer Says Its Antiviral Pill Is Highly Effective in
Treating Covid," *New York Times*, November 5, 2021, https://www.nytimes
.com/2021/11/05/health/pfizer-covid-pill.html?utm_source=Nature
+Briefing&utm_campaign=644cc12982-briefing-dy-20211108&utm
_medium=email&utm_term=0_c9dfd39373-644cc12982-41995895.

24. Brandon Malone and Elizabeth A. Campbell, "Molnupiravir: Coding
for Catastrophe," *Nature Structural & Molecular Biology*, 28 (2021):
706–8, https://www.nature.com/articles/s41594-021-00657-8.

25. Kathy Katella, "13 Things To Know About Paxlovid, the Latest COVID-19
Pill," *Yale Medicine, News*, November 29, 2022, https://www.yalemedicine
.org/news/13-things-to-know-paxlovid-covid-19.

26. Jim Reed, "Molnupiravir: First Pill to Treat Covid Gets Approval in
UK," BBC World News, November 4, 2021, https://www.bbc.com/news
/health-59163899.

27. Jason Kelly, presentation at Ferment Conference 2021, Boston, MA,
October 28, 2021.

28. "CDC Museum COVID-19 Timeline," CDC, updated August 16, 2022,
https://www.cdc.gov/museum/timeline/covid19.html#:~:text=January
%2020%2C%202020,January%2018%20in%20Washington%20state.

29. "COVID-19: How COVID-19 Vaccines Get to You," CDC, updated
April 19, 2021, https://www.cdc.gov/coronavirus/2019-ncov/vaccines
/distributing.html.

30. Charlotte Peet, "'Out of Control': Brazil's COVID Surge Sparks Regional
Fears," *Aljazeera News*, April 7, 2021, https://www.aljazeera.com/news
/2021/4/10/out-of-control-brazils-covid-surge-sparks-regional-fears;
"India's COVID-19 Crisis Has Spiralled Out of Control," *Economist*,
May 3, 2021, https://www.economist.com/graphic-detail/2021/05/03
/indias-covid-19-crisis-has-spiralled-out-of-control.

31. Poor People's Campaign, "Mapping the Intersections of Poverty,
Race, and COVID-19," Poor People's Campaign Report, April 2022,

https://www.poorpeoplescampaign.org/wp-content/uploads/2022/04/PoorPeoplesPandemicReport-Executive-Summary-April2022.pdf.

32. Matthew Fox, "Moderna, BioNTech Plunge as Bernie Sanders Slams 'Obscene' Profits and Tells Drug Makers to Share COVID-19 Vaccines with the World," Markets Insider, December 6, 2021, https://markets.businessinsider.com/news/stocks/moderna-biontech-stock-price-bernie-sanders-tweet-control-greed-vaccines-2021-12.

33. "South Africa's Biovac to Start Making Pfizer-BioNTech COVID-19 Vaccine in Early 2022—Exec," Reuters, December 6, 2021, https://www.reuters.com/business/healthcare-pharmaceuticals/south-africas-biovac-start-making-pfizer-biontech-covid-19-vaccine-early-2022-2021-12-06/.

34. "'Absolutely Unacceptable' COVID-19 Vaccination Rates in Developing Countries," World Bank, August 3, 2021, https://www.worldbank.org/en/news/podcast/2021/07/30/-absolutely-unacceptable-vaccination-rates-in-developing-countries-the-development-podcast.

35. Peter Schacknow, "Stocks Making the Biggest Moves Premarket: Moderna, Royal Caribbean, Cerner and Others," CNBC, December 20, 2021, https://www.cnbc.com/2021/12/20/stocks-making-the-biggest-moves-premarket-moderna-royal-caribbean-cerner-and-others.html.

36. "Bloomberg Billionaires Index: #296 Stéphane Bancel," Bloomberg, https://www.bloomberg.com/billionaires/profiles/stephane-bancel/ (accessed December 8, 2021).

37. Olivia Goldhill, "'We're Being Left Behind:' Rural Hospitals Can't Afford Ultra-Cold Freezers To Store the Leading COVID-19 Vaccine," Stat, November 11, 2020, https://www.statnews.com/2020/11/11/rural-hospitals-cant-afford-freezers-to-store-pfizer-covid19-vaccine/.

38. Nathan Vardi, "Ugur Sahin Becomes a Billionaire on Hopes for Technology Behind COVID-19 Vaccine," Forbes, June 1, 2020, https://www.forbes.com/sites/nathanvardi/2020/06/01/ugur-sahin-becomes-a-billionaire-on-hopes-for-technology-behind-Covid-19-vaccine/.

39. Rebecca Spalding and Joshua Franklin, "Germany's BioNTech Raises $150 Million in Smaller-than-Planned U.S. IPO Amid Market Volatility," Reuters, October 9, 2019, https://www.reuters.com/article/us-biontech-ipo/germanys-biontech-raises-150-million-in-smaller-than-planned-u-s-ipo-amid-market-volatility-idUSKBN1WO29B.

40. Emily Leproust (@EmilyLeproust), tweet, May 17, 2021, https://twitter
    .com/EmilyLeproust/status/1394353518811680776.
41. Jason Kelly, e-mail to investors, July 25, 2022.
42. Pieter Cullis, quoted in Rowland, "Inside Pfizer's Race."
43. "Dr. Pieter Cullis Talks Lipid Nanoparticles and Vaccines of the Future
    on CBC Radio's Quirks & Quarks," Life Sciences Institute, University
    of British Columbia, June 13, 2021, https://lsi.ubc.ca/2021/06/13/dr
    -pieter-cullis-talks-lipid-nanoparticles-and-vaccines-of-the-future
    -on-cbc-radios-quirks-quarks/.
44. David Meyer, "Pfizer Partner BioNTech Unveils Container-Based Vaccine
    Factories That Could Start Manufacturing Vaccine Doses in Africa This
    Year," *Fortune*, February 16, 2022, https://fortune.com/2022/02/16/pfizer
    -biontech-covid-vaccine-biontainers-inequality-africa-afrigen-who.
45. Christina Agapakis (@thisischristina), Tweet, September 16, 2021,
    https://mobile.twitter.com/thisischristina/status/1438469334473523203.

13. THE MOIRAI'S GIFT

1. Stephanie Nolen and Sheryl Gay Stolberg, "Pressure Grows on US
   Companies to Share Covid Vaccine Technology," *New York Times*,
   September 23, 2021, https://www.nytimes.com/2021/09/22/us/politics
   /covid-vaccine-moderna-global.html.
2. Massimo Florio, "Covid Vaccines Are a Very Expensive Miracle,"
   Research Professional News, December 2, 2021, https://www.research
   professionalnews.com/rr-news-europe-views-of-europe-2021-12
   -covid-vaccines-are-a-very-expensive-miracle/.
3. Margaret Talbot, "The Rogue Experimenters," *New Yorker*, May 18,
   2020, https://www.newyorker.com/magazine/2020/05/25/the-rogue
   -experimenters.
4. Lynn Yarris, "Microbial-Based Antimalarial Drug Shipped to Africa,"
   University of California, August 13, 2014, https://www.universityof
   california.edu/news/microbial-based-antimalarial-drug-shipped-africa.
5. "WHO Issues New Recommendations on Human Genome Editing
   for the Advancement of Public Health," World Health Organization,
   July 12, 2021, https://www.who.int/news/item/12-07-2021-who-issues
   -new-recommendations-on-human-genome-editing-for-the-advancement
   -of-public-health.

6. Maxine Singer, quoted in Katja Grace, *The Asilomar Conference: A Case Study in Risk Mitigation*, Machine Intelligence Research Institute Technical Report 2015-9 (Berkeley, CA: Machine Intelligence Research Institute, July 15, 2015), https://intelligence.org/files/TheAsilomar Conference.pdf.

7. Paul Berg, David Baltimore, Sydney Brenner, et al., "Asilomar Conference on Recombinant DNA Molecules," *Science* 188, no. 4192 (June 6, 1975): 991–94, https://www.science.org/doi/10.1126/science .1056638.

8. Katherine Andrews, "The Dark History of Forced Sterilization of Latina Women," Panoramas, University of Pittsburgh, October 30, 2017, https:// www.panoramas.pitt.edu/health-and-society/dark-history-forced -sterilization-latina-women.

9. Paul Berg, "Asilomar 1975: DNA Modification Secured," *Nature* 455 (September 2008): 290–91, https://doi.org/10.1038/455290a.

10. Berg, "Asilomar 1975," 291.

11. Recombinant DNA Advisory Committee Archives, NIH, https://osp .od.nih.gov/biotechnology/recombinant-dna-advisory-committee / (accessed November 9, 2022).

12. Michael Crichton, *Jurassic Park* (New York: Random House, 1990), 2.

13. Presidential Commission on the Study of Bioethical Issues, "New Directions: the Ethics of Synthetic Biology and Emerging Tech-nologies," Washington, DC, December 2010, https://bioethicsarchive .georgetown.edu/pcsbi/sites/default/files/PCSBI-Synthetic-Biology -Report-12.16.10_0.pdf.

14. Gigi Kwik Gronvall, "US Competitiveness in Synthetic Biology," *Health Security* 13, no 6 (December 1, 2015): 378–89, https://www.ncbi .nlm.nih.gov/pmc/articles/PMC4685481/.

15. T. Londrain, M. Meyer, and A. Perez, "Do-It-Yourself Biology: Chal-lenges and Promises for an Open Science and Technology Movement," *Systems and Synthetic Biology* 7 (2013):115–26, https://doi.org/10.1007 /s11693-013-9116-4.

16. Jean Peccoud, quoted in Talbot, "Rogue Experimenters."

17. Josiah Zayner, quoted in Tom Ireland, "I Want to Help Humans Genet-ically Modify Themselves," *The Guardian*, December 24, 2017, https:// www.theguardian.com/science/2017/dec/24/josiah-zayner-diy-gene -editing-therapy-crispr-interview.

18. Will Canine, author interview via Zoom, November 4, 2021.

19. Gregg Gonsalves, quoted in Talbot, "Rogue Experimenters."

20. William K. Bleser, Humphrey Shen, Hannah L. Crook, et al., "Pandemic-Driven Health Policies to Address Social Needs and Health Equity," *Health Affairs*, March 10, 2022, https://www.healthaffairs.org /do/10.1377/hpb20220210.360906.

21. Denis Rebrikov, quoted in David Cyranowski, "Russian 'CRISPR-Baby' Scientist Has Started Editing Genes in Human Eggs with Goal of Altering Deaf Gene," *Nature*, October 18, 2019, https://www.nature .com/articles/d41586-019-03018-0.

22. Facebook Group, Sickle Cell Disease Association of America, Inc., https://www.facebook.com/sicklecellcampaign/.

23 Heidi Ledford, "Beyond CRISPR Babies: How Human Genome Editing is Moving On After Scandal," *Nature*, (March 2, 2023). https://www .nature.com/articles/d41586-023-00625-w

24. "Cell Therapy Ethics," Novo Nordisk, https://www.novonordisk .com/science-and-technology/bioethics/stem-cell-ethics.html (accessed November 9, 2022).

25. D. Kriebel, J. Tickner, P. Epstein, et al., "The Precautionary Principle in Environmental Science," *Environmental Health Perspectives* 109, no. 9 (September 2001): 871–76, https://www.ncbi.nlm.nih.gov/pmc/articles /PMC1240435/.

26. G.J. Annas, et al., "A Code of Ethics for Gene Drive Research," *The CRISPR Journal* 4, no. 1 (February 2021): 19–24, https://www.liebertpub .com/doi/epdf/10.1089/crispr.2020.0096.

27. Maria Cohut, "Modified Herpes Virus Effective Against Late-Stage Melanoma," *Medical News Today*, January 26, 2019, https://www.medical newstoday.com/articles/324293.

28. "About the National Biodefense Science Board," U.S. Department of Health and Human Services, last updated December 17, 2020, https:// www.phe.gov/Preparedness/legal/boards/nbsb/Pages/about.aspx.

29. Amber Dance, "The Shifting Sands of Gain-of-Function Research," *Nature*, October 27, 2021, https://www.nature.com/articles/d41586-021-02903-x.

30. Nidhi Subbaraman, "US Officials Revisit Rules for Disclosing Risky Disease Experiments," Nature, January 27, 2020, https://www.nature .com/articles/d41586-020-00210-5.

31. Elizabeth Vitalis, author interview via Zoom, August 24, 2022.

32. Max Bazeman and Paresh Patel, "SPAC's: What You Need to Know," *Harvard Business Review*, July/August 2021, https://hbr.org/2021/07 /spacs-what-you-need-to-know.

33. Harry Sloan, quoted in Jim Lane, "Ginkgo's Insanely Great $15B Deal, the Who and How Behind Biggest Sector Deal Ever," *The Digest*, May 11, 2021, https://www.biofuelsdigest.com/bdigest/2021/05/11/ginkgos -insanely-great-15b-deal-the-who-how-behind-biggest-sector-deal-ever/.

34. Sophia Roosth, "Biology's Sharp Left Turn," *Grow*, no. 3, 2021, 92–101, https://www.growbyginkgo.com/2022/03/31/biologys-sharp-left-turn/.

35. Xavier Becerra and Antony Blinken, "Strengthening Global Health Security and Reforming International Health Regulations," *Journal of the American Medical Association* 326, no. 13 (August 2021): 1255–56, https://jamanetwork.com/searchresults?author=Xavier+Becerra&q=Xa vier+Becerra.

36. Editorial review by Donna Walton, quoted on Amazon site of Francoise Baylis, *Altered Inheritance* (Cambridge, MA: Harvard University Press, 2019).

37. Laurie Zoloth, "The Ethical Scientist in a Time of Uncertainty," *Cell* 184, no. 6 (March 2021): 1430–39, https://www.cell.com/cell/pdf/S0092 -8674(21)00240-3.pdf.

38. "Protecting Critical and Emerging Technologies from Foreign Threats," National Counterintelligence and Security Center, October 22, 2021, https://www.dni.gov/files/NCSC/documents/SafeguardingOurFuture /FINAL_NCSC_Emerging%20Technologies_Factsheet_10_22_2021.pdf.

39. Amy Maxmen and Noah Baker, "Coronapod: The Inequality at the Heart of the Pandemic," *Nature* Podcast, April 30, 2021 [audio], https:// doi.org/10.1038/d41586-021-01190-w.

40. "A Letter to Dr. Eric S. Lander, the President's Science Advisor and Nominee as Director of the Office of Science and Technology Policy," White House, January 20, 2021, https://www.whitehouse.gov /briefing-room/statements-releases/2021/01/20/a-letter-to-dr-eric-s -lander-the-presidents-science-advisor-and-nominee-as-director-of -the-office-of-science-and-technology-policy/.

41. "Ginkgo Bioworks Reports Fourth Quarter and Full Year 2022 Financial Results" March 1, 2023 https://www.prnewswire.com/news-releases /ginkgo-bioworks-reports-fourth-quarter-and-full-year-2022-financial -results-301760050.html.

42. Sheryl Gay Stolberg, "The U.S. Aims to Lift Covid Vaccine Manufacturing to Create a Billion Doses a Year," *New York Times*, November 17, 2021, https://www.nytimes.com/2021/11/17/us/covid-vaccines-supply.html.

43. White House, "Executive Order on Advancing Biotechnology and Biomanufacturing Innovation for a Sustainable, Safe and Secure American Bioeconomy," September 22, 2022, https://www.whitehouse .gov/briefing-room/presidential-actions/2022/09/12/executive-order -on-advancing-biotechnology-and-biomanufacturing-innovation-for -a-sustainable-safe-and-secure-american-bioeconomy.

14. TO THE PLANETS AND BEYOND: SYNTHETIC BIOLOGY IN SPACE

1. Kate Adamala, author interview by phone, July 11, 2019.

2. James Bevington, author interview via Zoom, August 17, 2020.

3. Biography of Lynn J. Rothschild, NASA, https://www.nasa.gov/content /lynn-j-rothschild (accessed November 9, 2022).

4. Kate Adamala, author interview via Zoom, July 10, 2021.

5. Adam Arkin, author interview via Zoom, June 30, 2021.

6. Jack Williamson, "Collision Orbit," in *Astounding Science Fiction*, July 1942.

7. H. G. Wells, *War of the Worlds* (London and New York: Harper & Brothers, 1898), 99.

8. Arthur C. Clarke, *2001: A Space Odyssey* (New York: New American Library, 1968).

9. C. McKay, "Terraforming Mars," *Journal of the British Interplanetary Society* 13, no. 204 (October 1982): 427–33; C. P. McKay, O. B. Toon, and J. F. Kasting, "Making Mars Habitable," *Nature* 352 (1991): 489–96; C. P. McKay and Margarita M. Marinova, "The Physics, Biology, and Environmental Ethics of Making Mars Habitable," *Astrobiology* 1, no. 1 (November 1, 2001): 89–110, https://www.ifa.hawaii.edu/~meech/a281 /handouts/McKay_astrobio01.pdf.

10. C. Verseux, I. Paulino-Lima, M. Baqué, et al., "Synthetic Biology for Space Exploration: Promises and Societal Implications," in *Ambivalences of Creating Life*, ed. K. Hagen, M. Engelhard, and G. Toepfer (Cham, Switzerland: Springer, 2016): 73–83, https://link.springer.com /chapter/10.1007/978-3-319-21088-9_4.

11. Adamala, author interview, by phone, July 10, 2019.

12. E. Boyden, et al., "Spatial Multiplexing of Fluorescent Reporters for Imaging Signalling Network Dynamics, *Cell* 183, no. 6 (2020): 1682–1698.e24.

13. Ike Swetlitz, "From Chemicals to Life: Scientists Try to Build Cells from Scratch," STAT, July 28, 2017, https://www.statnews.com/2017/07/28 /cell-build-from-scratch/.

14. John Cumbers, "NASA Is Quietly Funding This Project to Understand How Synthetic Biology Can Support Human Life in Space," *Forbes*, March 5, 2020, https://www.forbes.com/sites/johncumbers/2020/03/05 /nasa-is-quietly-funding-this-project-to-understand-how-synthetic -biology-can-support-human-life-in-space/?sh=13830ba26894.

15. Lynn Rothschild, author interview via Zoom, December 10, 2021.

16. "Space Mining Kits Blast Off for Tests in Orbit," University of Edinburgh, July 29, 2012, www.eurekalert.org/pub_releases/2019-07.

17. Lynn Rothschild, author interview via Microsoft Teams, December 20, 2021.

18. "Interview with Lynn Rothschild from Planetary Systems Branch," NASA, March 12, 2019, https://www.nasa.gov/ames/spacescience-and -astrobiology/interview-with-lynn-rothschild-from-planetary-systems -branch.

19. Rothschild, author interview.

20. Frank Tavares, "Could Future Homes on the Moon or Mars Be Made of Fungi?" NASA, January 14, 2020, https://www.nasa.gov/feature/ames /myco-architecture.

21. J. Urbina, A. Patil, K. Fujishima, et al., "A New Approach to Biomining: Bioengineering Surfaces for Metal Recovery from Aqueous Solutions," *Nature* 9 (November 2019): 16422, https://www.nature.com/articles /s41598-019-52778-2.

22. American Chemical Society, Press Release, "Space-Grown Lettuce Could Help Astronauts Avoid Bone Loss," March 22, 2022, https://www .acs.org/pressroom/newsreleases/2022/march/space-grown-lettuce -could-help-astronauts-avoid-bone-loss.html.

23. Lisa Nip, "How Humans Could Evolve to Survive in Space," TEDxBeaconStreet, March 30, 2016 [video], 8:13–10:03, https://www .ted.com/talks/lisa_nip_how_humans_could_evolve_to_survive _in_space?utm_campaign=tedspread&utm_medium=referral&utm _source=tedcomshare.

24. Lynn Rothschild, "Designing Nature," The Conference, September 7, 2018, https://videos.theconference.se/dr-lynn-rothschild-designing -nature.

25. J. Ehrlich, G. D. Massa, R. Wheeler, et al., "Plant Growth Optimization by Vegetable Production System in HI-SEAS Analog Habitat," presented at the AIAA SPACE and Astronautics Forum and Exposition, September 2017, https://www.researchgate.net/publication/319870597 _Plant_Growth_Optimization_by_Vegetable_Production_System _in_HI-SEAS_Analog_Habitat; Molly Kearns, "Inside the Mind of a Young Professional: A Conversation with Josh Ehrlich," Space Times, November 20, 2019, https://astronautical.org/2019/11/14/inside-the -mind-of-a-young-professional-a-conversation-with-josh-ehrlich/; Joshua Ehrlich, author interview via Zoom, July 10, 2021.

26. Aaron Pressman, "A Watershed Moment: With Public Debuts of Toast and Ginkgo, Boston Is Remaking Its Startup Scene," *Boston Globe*, September 22, 2021, https://www.bostonglobe.com/2021/09/22/business /watershed-moment-with-public-debuts-toast-ginkgo-boston-is -remaking-its-startup-scene/.

27. "Investors: Why Invest," Ginkgo Bioworks, https://investors.ginkgobio works.com/why-invest/default.aspx (accessed November 9, 2022).

28. Sophia Roosth, *Synthetic* (Chicago: University of Chicago Press, 2017).

29. Jean Peccoud, author interview via Zoom, October 20, 2021.

15. FUTURAMA

1. Alison Snyder, "Engineering an Economy Built by Biology," Axios, April 14, 2022, https://www.axios.com/engineering-bioeconomy-jobs -8940b7d7-c29d-4d71-a8b2-5ceef225ca02.html.

2. "The U.S. Bioeconomy: Charting a Course for a Resilient and Competitive Future," Schmidt Futures, April 2022, https://www.schmidtfutures .com/wp-content/uploads/2022/04/Bioeconomy-Task-Force-Strategy -4.14.22.pdf.

3. Amy Maxmen, "Unseating Big Pharma: The Radical Plan for Vaccine Equity," *Nature* 607, no. 7918 (July 13, 2022): 226–33, https://www.nature .com/immersive/d41586-022-01898-3/index.html.

4. "2022 iGEM Grand Jamboree," October 28, 2022, https://jamboree .igem.org/2022.

5. Megan Poinski, "Believer Meats Breaks Ground On World's Largest Cultivated Meat Plant," *Food Dive*, December 8, 2022, https://www.fooddive.com/news/believer-meats-commercial-scale-cultivated-meat-groundbreaking-future-meat-technologies/638263.

6. Some economic historians disagree about the number of industrial revolutions, whether three or four. For my purpose, in trying to place synthetic biology and gene editing in a pantheon of industrial history, a larger figure felt more logical. On the revolutionary nature of synthetic biology, see Jean Peccoud, "Synthetic Biology: Fostering the Cyber-biological Revolution," *Synthetic Biology* 1, no. 1 (May 27, 2016): ysw001, https://academic.oup.com/synbio/article/1/1/ysw001/2543666.

7. Jason D. Rowley, "DCVC to Raise $250 Million for Bio-Focused Fund, Incorporating Monsanto Leadership, *Crunchbase News*, April 20, 2018, https://news.crunchbase.com/venture/dcvc-raise-250-million-bio-focused-fund-incorporating-monsanto-leadership/.

8. Thomas Ybert, "DNA Printing Evolves," Technology Networks: Genomics Research, April 19, 2022, https://www.technologynetworks.com/genomics/articles/dna-printing-evolves-360154.

9. David Kirk, "A Game of Genomes: Synthetic Biology in the United Kingdom," SynBioBeta, July 23, 2019, https://synbiobeta.com/a-game-of-genomes-uk-synthetic-biology-research-centers/.

10. "The U.S. Bioeconomy," Schmidt Futures, 47.

11. John Cumbers, "Trade Deal or Not, China Invests Big in Synthetic Biology," *Forbes*, August 26, 2019, https://www.forbes.com/sites/johncumbers/2019/08/26/trade-deal-or-no-china-is-investing-big-in-synthetic-biology/?sh=61e3559f2e58.

12. White House, Executive Order on Advancing Biotechnology and Bio-manufacturing Innovation for a Sustainable, Safe, and Secure American Bioeconomy, September 12, 2022.

13. Andrew Bary, "Synthetic Biology Could Be the Next Big Thing. Here's How to Play It," *Barron's*, July 16, 2021, https://www.barrons.com/articles/biotech-stocks-amyris-zymergen-gingko-bioworks-soaring-eagle-acquisition-51626478742.

14. Simply Wall Street, "Amyris, Inc. (NASDAQ: AMRS): Are Analysts Optimistic?," October 13, 2022, https://simplywall.st/stocks/us/materials/nasdaq-amrs/amyris/news/amyris-inc-nasdaqamrs-are-analysts-optimistic.

15. Dan Widmaier, author interview via Zoom, December 13, 2021.

16. Eben Bayer, "For Synthetic Biology to Scale, It Has to Think Outside the Tank," *Forbes*, July 25, 2022, https://www.forbes.com/sites/ebenbayer/2022/07/25/for-bio-manufacturing-to-scale-it-has-to-think-outside-the-tank/.

17. Author visit to LanzaTech, November 10, 2021.

18. Peter Forbes, "Virtuous Cycles," *New Humanist*, July 24, 2022, https://newhumanist.org.uk/articles/5978/virtuous-cycles.

19. Tara Deans, author interview, Salt Lake City, Utah, July 13, 2021.

20. Sophia Roosth, "Biology's Sharp Left Turn," *Grow*, no. 3, 2021, 92–97.

21. Christina Agapakis, author interview via Zoom, July 7, 2022.

22. Rebecca Mackelprang, Emily R. Aurand, Roel A. L. Bovenberg, et al., "Guiding Ethical Principles in Engineering Biology," *ACS Synthetic Biology*, 10 (2021): 907–10, https://pubs.acs.org/doi/pdf/10.1021/acssynbio.1c00129.

23. Maxmen, "Unseating Big Pharma," 227.

24. Michael Jewett, author interview via Zoom, June 21, 2022.

25. Jay Keasling, "Agriculture and Synthetic Biology," March 25, 2021, Built With Biology Presentation on Zoom, URL no longer available.

INDEX